Modern Blast Furnace Ironmaking
an introduction

Maarten Geerdes
Hisko Toxopeus
Cor van der Vliet

Modern Blast Furnace Ironmaking
an introduction

With contributions from
Renard Chaigneau
Tim Vander
Jennifer Wise

Second Edition, 2009

© 2009 The authors and IOS Press. All rights reserved.

ISBN 978-1-60750-040-7

Published by IOS Press under the imprint Delft University Press

Publisher
IOS Press BV
Nieuwe Hemweg 6b
1013 BG Amsterdam
The Netherlands
tel: +31-20-688 3355
fax: +31-20-687 0019
email: info@iospress.nl
www.iospress.nl
www.dupress.nl

LEGAL NOTICE
The publisher is not responsible for the use which might be made of the following information.

PRINTED IN THE NETHERLANDS

Preface

In the second edition of "Modern Blast Furnace Ironmaking", we have included our insights gained during numerous discussions with colleagues all over the world and our own internal core team. We have also greatly benefited from the many courses and questions raised by the participants in these courses.

The objective of this book is to share our insights that optimization of the blast furnace is not only based on "best practice transfer", but also requires conceptual understanding why a measure works well in some cases and does not work in other situations. In other words, operational improvement is not only based on know–how, but on know–why as well.

We are indebted to many people we have worked with. We are grateful for the contributions of Renard Chaigneau, Tim Vander and Jennifer Wise, who re–wrote chapters III, IV and X respectively. Ing. Oscar Lingiardi, Prof. Dr. Fernando Tadeu Pereira de Medeiros, Prof. Dr. I. Kurunov and Ing. Vincenzo Dimastromatteo have given us valuable comment and taken care of translations into Spanish, Portuguese, Russian and Italian. Ternium Siderar provided the examples of data presentation at an operating blast furnace. A special word of thanks to John Ricketts, who helped develop the material covered in the first edition of this book into a blast furnace operator course, has helped enormously with teaching materials and has shared his insights with us for more than 15 years.

Danieli Corus directors Mr. P. Zonneveld and previously N. Bleijendaal encouraged us to write the first edition of this book. We thank Edo Engel at XLmedia for the editing.

We learn by sharing our knowledge. We wish the same to our readers.

Maarten Geerdes, Hisko Toxopeus, Cor van der Vliet,

IJmuiden, July 2009

Contents

Preface		v
Contents		vii
List of Symbols and Abbreviations		xi
Chapter I	Introduction of the Blast Furnace Process	1
1.1	What is driving the furnace?	4
1.2	The equipment	6
1.3	Book overview	10
Chapter II	The Blast Furnace: Contents and Gas Flow	11
2.1	The generation of gas and gas flow through the burden	11
2.2	Furnace efficiency	15
2.3	An example of gas flow and contents of a blast furnace	16
Chapter III	The Ore Burden: Sinter, Pellets, Lump Ore	19
3.1	Introduction	19
3.2	Iron ore	20
3.3	Quality demands for the blast furnace burden	22
3.4	Sinter	26
3.5	Pellets	30
3.6	Lump ore	34
3.7	Interaction of burden components	35
Chapter IV	Coke	37
4.1	Introduction: function of coke in the blast furnace	37
4.2	Coal blends for coke making	38
4.3	Coke quality concept	39
4.4	Coke size distribution	43
4.5	Mechanical strength of coke	44
4.6	Overview of international quality parameters	46
Chapter V	Injection of Coal, Oil and Gas	47
5.1	Coal injection: equipment	48
5.2	Coal specification for PCI	49
5.3	Coal injection in the tuyeres	51
5.4	Process control with pulverised coal injection	52
5.5	Circumferential symmetry of injection	56
5.6	Gas and oil injectants	57

Chapter VI	Burden Calculation and Mass Balances	59
6.1	Introduction	59
6.2	Burden calculation: starting points	59
6.3	An example of a burden calculation	60
6.4	Process calculations: a simplified mass balance	61
Chapter VII	The Process: Burden Descent and Gas Flow Control	67
7.1	Burden descent: where is voidage created?	67
7.2	Burden descent: system of vertical forces	69
7.3	Gas flow in the blast furnace	71
7.4	Fluidisation and channelling	78
7.5	Burden distribution	78
7.6	Coke layer	84
7.7	Ore layer thickness	85
7.8	Erratic burden descent and gas flow	88
7.9	Blast furnace instrumentation	90
7.10	Blast furnace daily operational control	90
Chapter VIII	Blast Furnace Productivity and Efficiency	93
8.1	The raceway	93
8.2	Carbon and iron oxides	96
8.3	Temperature profile	104
8.4	What happens with the gas in the burden?	104
8.5	Oxygen and productivity	106
8.6	Use of metallic iron	107
8.7	How iron ore melts	107
8.8	Circumferential symmetry and direct reduction	112
Chapter IX	Hot Metal and Slag	115
9.1	Hot metal and the steel plant	115
9.2	Hot metal composition	116
9.3	Silicon reduction	117
9.4	Hot metal sulphur	118
9.5	Slag	118
9.6	Hot metal and slag interactions: special situations	122
Chapter X	Casthouse Operation	125
10.1	Objectives	125
10.2	Liquid iron and slag in the hearth	125
10.3	Removal of liquids through the taphole	127
10.4	Typical casting regimes	128
10.5	Taphole drill and clay gun	130
10.6	Hearth liquid level	131
10.7	Delayed casting	132
10.8	No slag casting	134
10.9	One–side casting	135
10.10	Not dry casts	137

10.11	Defining a dry hearth	139
10.12	Oxygen lancing	139
10.13	Cast data recording	140
Chapter XI	Special Situations	141
11.1	Fines in ore burden	141
11.2	Moisture input	143
11.3	Recirculating elements	144
11.4	Charging rate variability	145
11.5	Stops and start–ups	145
11.6	Blow–down	147
11.7	Blow–in from new	148
Glossary		151
Annex I	Further Reading	153
Annex II	References	154
Annex III	Rules of Thumb	156
Annex IV	Coke Quality Tests	157
Index		161

List of Symbols and Abbreviations

B2, B3, B4	basicity, ratio of two, three or four components
bar	pressure, atmosphere relative
°C	degrees centigrade
C	carbon
cm	centimetre
CO	carbon monoxide
CO_2	carbondioxide
CRI	coke reactivity index
CSR	coke strength after reaction
Fe	iron
GJ	giga joule
H_2	hydrogen
H_2O	water
HGI	hard grove index
HMS	harmonic mean size
HOSIM	hoogovens simulatie (blast furnace simulation)
HV	high volatile
IISI	International Iron & Steel Institute
ISO	International Organisation for Standardization
JIS	Japanese Industrial Standard
K	potassium
kg	kilogram
kmole	kilomole
LV	low volatile
m^3 STP	cubic metre at standard temperature and pressure
mm	millimetre
Mn	manganese
Mt	million ton
N_2	nitrogen
Na	sodium
O_2	oxygen
P	phosphorous
PCI	pulverised coal injection
RAFT	raceway adiabatic flame temperature
RR	replacement ratio
s	second
S	sulphur
Si	silicon

Standard Coke	coke with 87.5 % carbon
STP	standard temperature and pressure
t	tonne (1000 kg)
tHM	tonne hot metal
Ti	titanium
VDEh	Verein Deutscher Eisenhüttenleute
VM	volatile matter

1 Introduction of the Blast Furnace Process

Two different process routes are available for the production of steel products, namely the blast furnace with oxygen steelmaking and the electric arc steelmaking route. The routes differ with respect to the type of products that can be made, as well as the raw materials used. The blast furnace–oxygen steelmaking route mainly produces flat products, while electric arc steelmaking is more focused on long products. The former uses coke and coal as the main reductant sources and sinter, pellets and lump ore as the iron–bearing component, while the latter uses electric energy to melt scrap. The current trend is for electric arc furnaces to be capable of also producing flat products. Nevertheless, the blast furnace–oxygen steelmaking route remains the primary source for worldwide steel production, as shown in Figure 1.1.

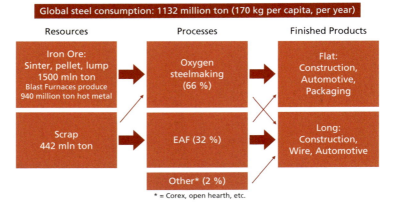

Figure 1.1   Steelmaking routes and raw materials
              (IISI Steel Statistical Yearbook and World Steel in Figures, 2007)

Hot metal is produced in a blast furnace, from where it is transported as liquid hot metal to the steel plant where refinement of hot metal to steel takes place by removing elements such as sulphur, silicon, carbon, manganese and phosphorous. Good performance of the steel plant requires consistent hot metal quality of a given specification. Typically the specification demands silicon content between 0.3 % and 0.7 %, manganese between 0.2 % and 0.4 %, phosphorous in the range 0.06–0.08 % or 0.1–0.13 % and a temperature as high as possible.

In the blast furnace process iron ore and reducing agents (coke, coal) are transformed to hot metal and slag is formed from the gangue of the ore burden and the ash of coke and coal. Hot metal and liquid slag do not mix and remain separate from each other with the slag floating on top of the denser iron. The iron can then be separated from the slag in the casthouse.

Let us now consider the contents of a blast furnace at any given moment. Ore and coke are charged in discrete layers at the top of the furnace. From studies of quenched furnaces it was evident that these layers of ore and coke remain until the temperatures are high enough for softening and melting of the ore to begin. Quenched furnaces are "frozen in action" with the help of water or nitrogen and examples of quenched blast furnaces as well as a solidiftied cohesive zone are presented in Figures 1.2a and 1.2b.

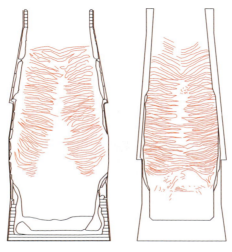

Figure 1.2a   Dissections of quenched blast furnaces Kakogawa 1 and Tsurumi (Based on Omori et al, 1987)

Figure 1.2b   Cohesive zone left after blow–down, courtesy J. Ricketts, ArcelorMittal

The quenched blast furnace shows clearly the layer structure of coke and ore. Further analysis reveals information about the heating and melting of the ore as well of the progress of chemical reactions.

As indicated in Figure 1.3, at any moment, an operating blast furnace contains, from top downwards: :
– Layers of ore and coke.
– An area where ore starts to soften and melt, known as the softening–melting zone.
– An area where there is only coke and liquid iron and slag, called the "active coke" or dripping zone.
– The dead man, which is a stable pile of coke in the hearth of the furnace.

A blast furnace has a typical conical shape. The sections from top down are:
– Throat, where the burden surface is.
– The stack, where the ores are heated and reduction starts.
– The bosh parallel or belly and
– The bosh, where the reduction is completed and the ores are melted down.
– The hearth, where the molten material is collected and is cast via the taphole.

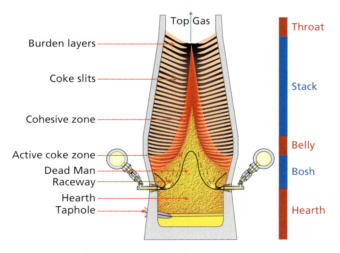

Figure 1.3   The zones in the blast furnace

1.1 What is driving the furnace?

1.1.1 Process description

The inputs and outputs of the furnace are given in Figure 1.4.

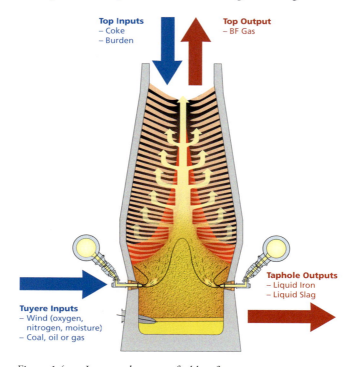

Figure 1.4   Input and output of a blast furnace

- A blast furnace is filled with alternating layers of coke and the iron ore–containing burden.
- Hot blast is blown into the blast furnace via tuyeres. A tuyere is a cooled copper conical pipe numbering up to 12 in smaller furnaces, and up to 42 in bigger furnaces through which pre–heated air (up to more than 1200 °C) is blown into the furnace.
- The hot blast gasifies the reductant components in the furnace, those being coke as well as auxiliary materials injected via the tuyeres. In this process, the oxygen in the blast is transformed into gaseous carbon monoxide. The resulting gas has a high flame temperature of between 2100 and 2300 °C . Coke in front of the tuyeres is consumed thus creating voidage The driving forces in the blast furnace are illustrated in Figure 1.5.
- The very hot gas ascends through the furnace , carrying out a number of vital functions.
- Heats up the coke in the bosh/belly area.

- Melting the iron ore in the burden, creating voidage.
- Heats up the material in the shaft zone of the furnace.
- Removes oxygen of the ore burden by chemical reactions.
- Upon melting, the iron ore produces hot metal and slag, which drips down through the coke zone to the hearth, from which it is removed by casting through the taphole. In the dripping zone the hot metal and slag consume coke, creating voidage. Additional coke is consumed for final reduction of iron oxide and carbon dissolves in the hot metal, which is called carburisation.

The blast furnace can be considered as a counter current heat and mass exchanger, as heat is transferred from the gas to the burden and oxygen from the burden to the gas. Gas ascends up the furnace while burden and coke descend down through the furnace. The counter current nature of the reactions makes the overall process an extremely efficient one.

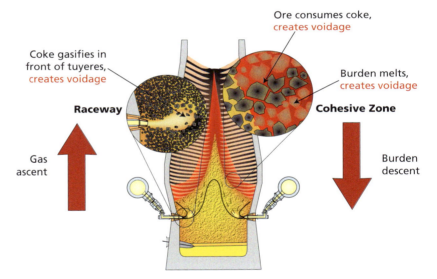

Figure 1.5 The driving force of a blast furnace: the counter current process creates voidage at the indicated areas causing the burden to descend

A typical example of the temperature profile in the blast furnace is shown in Figure 1.6. It is shown that the softening/melting zone is located in an area where temperatures are between 1100 and 1450 °C. The temperature differences in the furnace are large. In the example the temperature gradients are bigger in the horizontal direction than in the vertical direction, which will be explained in Chapter VI.

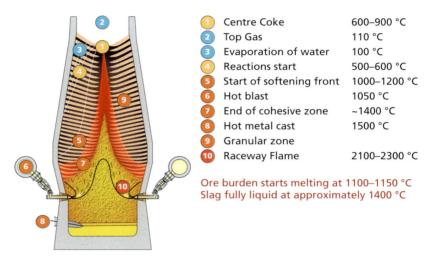

1	Centre Coke	600–900 °C
2	Top Gas	110 °C
3	Evaporation of water	100 °C
4	Reactions start	500–600 °C
5	Start of softening front	1000–1200 °C
6	Hot blast	1050 °C
7	End of cohesive zone	~1400 °C
8	Hot metal cast	1500 °C
9	Granular zone	
10	Raceway Flame	2100–2300 °C

Ore burden starts melting at 1100–1150 °C
Slag fully liquid at approximately 1400 °C

Figure 1.6   Temperature profile in a blast furnace (typical example)

1.2 The equipment

1.2.1 Equipment overview

An overview of the major equipment is shown in Figure 1.7. These include:
– Hot Blast Stoves. Air preheated to temperatures between 1000 and 1250 °C is produced in the hot blast stoves and is delivered to the furnace via a hot blast main, bustle pipe, tuyere stocks and finally through the tuyeres. The hot blast reacts with coke and injectants. The high gas speed forms the area known as the raceway in front of the tuyeres.
– Stock house. The burden materials and coke are delivered to a stock house. The materials are screened and then weighted before final delivery into the furnace. The stock house is operated automatically. Corrections for coke moisture are generally made automatically. The burden materials and coke are brought to the top of the furnace via skips or via a conveyor belt, where they are discharged into the furnace in separate layers of ore and coke.
– Gas cleaning. The top gas leaves the furnace via uptakes and a down–comer. The top gas will contain many fine particles and so to remove as many of these as possible the top gas is lead through a dust catcher and wet cleaning system.

- Casthouse. The liquid iron and slag collect in the hearth of the furnace, from where they are tapped via the taphole into the casthouse and to transport ladles.
- Slag granulation. The slag may be quenched with water to form granulated slag, which is used for cement manufacturing.

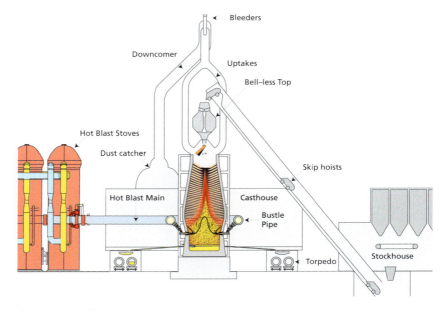

Figure 1.7   Blast furnace general arrangement

The top of the blast furnace is closed, as modern blast furnaces tend to operate with high top pressure. There are two different systems:
- The double bell system, often equipped with a movable throat armour.
- The bell less top, which allows easier burden distribution.
  Examples of both types are schematically shown in Figure 1.8.

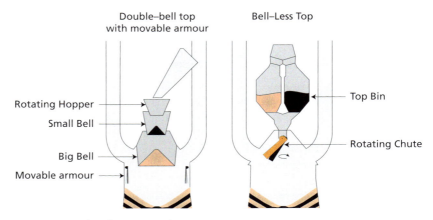

Figure 1.8   Blast furnace top charging systems

1.2.2 Blast furnace construction

There are basically two construction techniques to support blast furnaces. The classic design utilises a supported ring, or lintel at the bottom of the shaft, upon which the higher levels of the furnace rests. The other technique is a freestanding construction requiring an independent support for the blast furnace top and the gas system. The required expansion (thermal as well as from the pressure) for the installation is below the lintel that is in bosh/belly area for the lintel furnace, while the compensator for expansion in the freestanding furnace is at the top, as indicated in Figure 1.9.

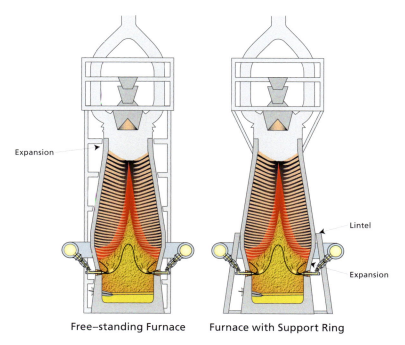

Figure 1.9   Blast furnace constructions

1.2.3 Blast furnace development

Blast furnaces have grown considerably in size during the 20th century. In the early days of the 20th century, blast furnaces had a hearth diameter of 4 to 5 metres and were producing around 100,000 tonnes hot metal per year, mostly from lump ore and coke. At the end of the 20th century the biggest blast furnaces had between 14 and 15 m hearth diameter and were producing 3 to 4 Mt per year.

The ore burden developed, so that presently high performance blast furnaces are fed with sinter and pellets. The lump ore percentage has generally decreased to 10 to 15 % or lower. The reductants used developed as well: from operation with coke only to the use of injectant through the tuyeres. Mainly oil injection in the 1960's, while since the early 1980's coal injection is used extensively. Presently, about 30 to 40 % of the earlier coke requirements have been replaced by injection of coal and sometimes oil and natural gas.

The size of a blast furnace is often expressed as its hearth diameter or as its "working volume" or "inner volume". The working volume is the volume of the blast furnace that is available for the process i.e. the volume between the tuyeres and the burden level. Definitions of working volume and inner volume are given in Figure 1.10.

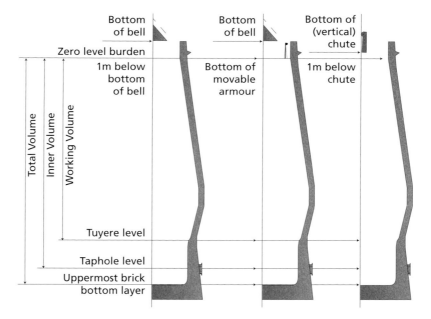

Figure 1.10   Definitions of working volume and inner volume

Presently, very big furnaces reach production levels of 12,000 t/d or more. E.g. the Oita blast furnace No. 2 (NSC) has a hearth diameter of 15.6 meter and a production capacity of 13,500 t/d. In Europe, the Thyssen–Krupp Schwelgern No. 2 furnace has a hearth diameter of 14.9 m and a daily production of 12,000 t/d.

1.3 Book overview

Blast furnace ironmaking can be discussed from 3 different perspectives:
- The operational approach: discussing the blast furnace with its operational challenges.
- The chemical technology approach: discussing the process from the perspective of the technologist who analyses progress of chemical reactions and heat and mass balances.
- The mechanical engineering approach focussing on equipment.

The focus of this book is the "operator's view", with the aim to understand what is going on inside the furnace. To this end the principles of the process are discussed (Chapter II) followed by the demands on burden quality (Chapter III) and coke and auxiliary reductants (Chapters IV and V). Simplified calculations of burden and top gas are made (Chapter VI). The control of the process is discussed in Chapter VII: burden descent and gas flow control. The issues pertinent to understanding the blast furnace productivity and efficiency are presented in Chapter VIII. Subsequently, hot metal and slag quality (Chapter IX), casthouse operation (Chapter X) and special operational conditions like stops and starts, high moisture input or high amounts of fines charged into the furnace (Chapter XI) are discussed.

II *The Blast Furnace: Contents and Gas Flow*

2.1 The generation of gas and gas flow through the burden

The blast furnace process starts when pre-heated air, or 'hot blast' is blown into the blast furnace via the tuyeres at a temperature of up to 1200 °C. The hot blast burns the fuel that is in front of the tuyere, which is either coke or another fuel that has been injected into the furnace through the tuyeres. This burning generates a very hot flame and is visible through the peepsites as the "raceway". At the same time the oxygen in the blast is transformed into gaseous carbon monoxide (CO). The resulting gas has a flame temperature of between 2000 and 2300 °C. The hot flame generates the heat required for melting the iron ore (Figure 2.1).

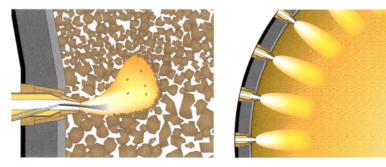

Figure 2.1   The raceway, horizontal and vertical sections

The blast furnace is a counter current reactor (Figure 2.2, next page). The driving force is the hot blast consuming coke at the tuyeres. In this chapter the gas flow through the furnace is analysed in more detail. The charge consists of alternating layers of ore burden (sinter, pellets, lump ore) and coke. The burden is charged cold and wet into the top of the furnace, while at the tuyeres the hot blast gasifies the hot coke. Towards the burden stockline (20 to 25 m from tuyeres to burden surface) the gas temperature drops from a flame temperature of 2200 °C to a top gas temperature of 100 to 150 °C.

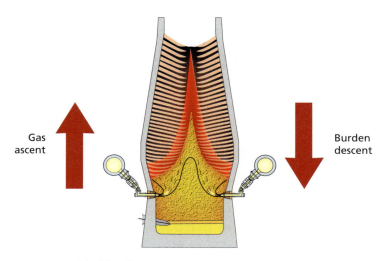

Figure 2.2   The blast furnace as a counter current reactor

The process starts with the hot blast through the tuyeres, which gasifies the coke and coal in the raceway (Figure 2.1). The reactions of the coke create hot gas, which is able to melt the ore burden. Consumption of coke and melting of the ore burden creates space inside the furnace, which is filled with descending burden and coke. The oxygen in the blast will gasify the coke to generate carbon monoxide (CO). For every molecule of oxygen 2 molecules of carbon monoxide are formed. If blast is enriched from its base level of 21 to 25 % oxygen, then every cubic meter (m^3 STP) oxygen will generate 2 m^3 STP of CO. So if the blast has 75 % of nitrogen and 25 % of oxygen, the bosh gas will consist of 60 % (i.e. 75/(75+2x25)) nitrogen and 40 % CO gas. In addition a huge amount of heat is generated in the raceway from the combustion of coke and coal (or oil, natural gas). The heat leads to a high flame temperature, which generally is in the range of 2000 to 2300 °C. Since this temperature is higher than the melting temperature of iron and slag, the heat in the hot gas can be used to melt the burden. Flame temperature is discussed in more detail in section 8.1.3.

The hot gas ascends through the ore and coke layers to the top of the furnace. If there was only coke in the blast furnace, the chemical composition of the gas would remain constant but the temperature of the gas would lower as it comes into contact with the colder coke layers high in the furnace. A presentation of the gas flowing through a blast furnace filled with coke is presented in Figure 2.3. To the experienced blast furnace operator the furnace filled with coke only may seem a theoretical concept. However, in some practical situations, like the blow–in of a new furnace or when taking a furnace out of operation for a long time (banking) the furnace is almost entirely filled with coke.

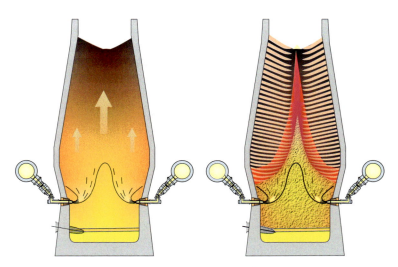

Figure 2.3   Gas flow in a furnace filled with coke only (left) and in a furnace filled with alternating layers of coke and ore (right).

In the normal operational situation the furnace is filled with alternating coke and ore layers. About 35 to 45 layers of ore separate the coke. It is important to note that the permeability of coke is much better than the permeability of ore (see also Figure 7.6). This is due to the fact that coke is much coarser than sinter and pellets and that the void fraction within the coke layer is higher. For example, the mean size of coke in a blast furnace is typically 45 to 55 mm, while the average size of sinter is 10 to 20 mm and of pellets is 10 to 12 mm. Consequently, the burden layers determine how the gas flows through the furnace, while the coke layers function as gas distributors.

If gas flows from the bosh upwards, what happens to the gas as it gradually cools down? Firstly, the heat with a temperature in excess of 1400 °C, the melting temperature of the slag, is transferred to the layered burden and coke, causing the metallic portion to melt. In the temperature range from 1400 to 1100 °C the burden will soften and stick together rather than melt. In the softening and melting zone the remaining oxygen in the ore burden is removed, which generates additional carbon monoxide. This is referred to as the direct reduction reaction (see section 7.2.1), which only occurs in the lower furnace.

The gas has now cooled to about 1100 °C and additional gas has been generated. Since the direct reduction reaction costs a lot of energy, the efficiency of the furnace is largely dependant on the amount of oxygen removed from the burden materials before reaching this 1100 °C temperature.

In summary:
- Heat is transferred from the gas to the ore burden, which melts and softens (over 1100 °C).
- Residual oxygen in the burden is removed and additional CO is generated. This is known as the direct reduction reaction.

Upon further cooling down the gas is capable of removing oxygen from the ore burden, while producing carbon dioxide (CO_2). The more oxygen that is removed, the more efficient the furnace is. Below temperatures of 1100 °C the following takes place:
- Heat is transferred from the gas to the burden.
- CO_2 gas is generated from CO gas, while reducing the amount of oxygen of the ore burden. This is called the gas reduction reaction, and in literature it is sometimes called "indirect reduction" as opposed to "direct reduction". No additional gas is generated during this reaction.
- A similar reaction takes place with hydrogen. Hydrogen can remove oxygen from the burden to form water (H_2O).

Higher in the furnace, the moisture in the burden and coke evaporates and so is eliminated from the burden before any chemical reactions take place.

If we follow the burden and coke on its way down the stack, the burden and coke are gradually heated up. Firstly the moisture is evaporated, and at around 500 °C the removal of oxygen begins. A simplified schedule of the removal of oxygen from the ore burden is shown in Figure 2.4.

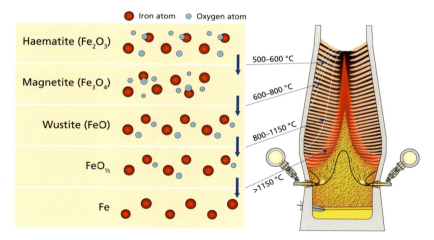

Figure 2.4 Schematic presentation of reduction of iron oxides and temperature

The first step is the reduction of haematite (Fe_2O_3) to magnetite (Fe_3O_4). The reduction reaction generates energy, so it helps to increase the temperature of the burden. In addition, the reduction reaction creates tension in the crystal structure of the burden material, which may cause the crystal structure to break into smaller particles. This property is called low–temperature disintegration. Several tests are available to quantify the effects (see Chapter III). Further down in the furnace the temperature of the burden increases gradually until the burden starts to soften and to melt in the cohesive zone. The molten iron and slag are collected in the hearth.

We now consider the interaction between the gas and the ore burden. The more the gas removes oxygen from the ore burden, the more efficient the blast furnace process is. Consequently, intimate contact between the gas and the ore burden is very important. To optimise this contact the permeability of the ore burden must be as high as possible. The ratio of the gas flowing through the ore burden and the amount of oxygen to be removed from the burden must also be in balance.

Experience has shown that many problems in the blast furnace are the consequence of low permeability ore layers. Therefore, the permeability of the ore layers across the diameter of the furnace is a major issue. The permeability of an ore layer is largely determined by the amount of fines (under 5 mm) in the layer. Generally, the majority of the fines are generated by sinter, if it is present in the charged burden or from lump ores. The problem with fines in the furnace is that they tend to concentrate in rings in the furnace. As fines are charged to the furnace they concentrate at the point of impact where the burden is charged. They are also generated by low temperature reduction– disintegration. Thus, it is important to screen the burden materials well, normally with 5 or 6 mm screens in the stock house, and to control the low temperature reduction–disintegration characteristics of the burden.

2.2 Furnace efficiency

The process efficiency of the blast furnace, generally considered to be the reductant rate per tonne hot metal, is continuously monitored through measurement of the chemical composition of the top gas. The efficiency is expressed as the gas utilisation, that is the percentage of the CO gas that has been transformed to CO_2, as defined in the following expression:

$$\eta_{CO} = \frac{CO_2}{(CO + CO_2)}$$

In addition, at modern furnaces the gas composition over the radius is frequently measured. The latter shows whether or not there is a good balance between the amount of reduction gas and the amount of ore in the burden. The wall zone is especially important and so the coke percentage in the wall area should not be too low. The wall area is the most difficult place to melt the burden as that is where the burden thickness is at it's highest across the radius, and also because the gas at the wall loses much of its temperature to cooling losses.

The top gas analysis gives a reasonably accurate indication of the efficiency of the furnace. When comparing different furnaces one should realise that the hydrogen also takes part in the reduction process (paragraph 7.2.4).

The gas utilisation also depends on the amount of oxygen that must be removed. Since pellets have about 1.5 atoms of oxygen per atom of Fe (Fe_2O_3) and sinter has about 1.45 (mix of Fe_2O_3 and Fe_3O_4), the top gas utilisation will be lower when using sinter. It can be calculated as about 2.5 % difference of the top gas utilisation, when comparing an all pellet burden with an all sinter burden.

2.3 An example of gas flow and contents of a blast furnace

The contents of a blast furnace can be derived from operational results. How long do the burden and gas reside within the furnace? Consider an example of a large, high productivity blast furnace with a 14 metre hearth diameter. It has a daily production of 10,000 t hot metal (tHM) at a coke rate of 300 kg/tHM and a coal injection rate of 200 kg/t. Moisture in blast and yield losses are neglected. Additional data is given in Table 2.1.

	Consumption		Specific weight		Carbon content	
Ore burden	1580	kg/tHM	1900	kg/m³		
Coke	300	kg/tHM	500	kg/m³	87	%
Coal	200	kg/tHM			78	%
Blast Volume	6500	m³ STP/min	1.3	kg/m³ STP		
Top Gas			1.35	kg/m³ STP		
O_2 in blast	25.6	%				
Working volume	3800	m³	(500 m³ used for active coke zone)			
Throat diameter	10	m				
A charge contains	94.8	t ore burden	18	t coke		
A ton hot metal contains	945	kg Fe	45	kg carbon		
Voidage in shaft	30	%				

Table 2.1 Data for calculation of blast furnace contents
1 tonne hot metal contains 945 kg Fe= 945/55.6 = 17.0 kmole

2.3.1 How much blast oxygen is used per tonne hot metal?

Oxygen from the blast volume amounts to 0.256 x 6500 m³ STP/min = 1664 m³ STP oxygen/min. The production rate is 10,000/(24x60) = 6.94 tHM/min. So the oxygen use is 1664/6.94 = 240 m³ STP blast oxygen/tHM.

2.3.2 How often are the furnace contents replaced?

To produce a tonne of hot metal, the furnace is charged with:
- 300 kg coke: 0.64 m³ (300/470) volume
- 1580 kg sinter/pellets: 0.88 m³ (1580/1800) volume
- Total per tonne of hot metal: 1.52 m³ volume

Production is 10,000 tonne per day, which is 10,000x1.52 m³ = 15,200 m³ volume per day. This material can be contained in the working volume of the furnace, with exception of the volume used for the active coke zone. So the contents of the furnace are refreshed 4.6 times per day (15,200/(3800–500)). This means the burden charged at the top reaches the tuyeres in 5.2 hours.

2.3.3 How many layers of ore are in the furnace at any moment?

The number of ore layers depends on the layer thickness or the weight of one layer in the burden. It can vary from furnace to furnace and depends on the type of burden used so there is a large variety of appropriate burden thicknesses. A typical range is 90–95 tonne of burden per layer. A layer contains 94.8 tonne, so about 60 tonne hot metal. In 5.2 hours, the furnace produces 2,167 tonne, which corresponds to 36 layers of ore (2167/60). In our example, taking a throat diameter of 10 m, the ore layer is 67 cm and the coke layer is an average of 49 cm at the throat.

2.3.4 What happens to the carbon of the coke and coal?

One tonne of HM requires:
- 300 kg coke, C content 87 %: 261 kg C
- 200 kg coal, C content 78 %: 160 kg C
- Total carbon: 417 kg C

About 45 kg carbon dissolves in the hot metal. The balance leaves the furnace through the top, which is 421–45 = 372 kg. It leaves the furnace as CO and CO_2.

2.3.5 Estimate how long the gas remains in the furnace

The blast volume is 6500 m³ STP/min with 25.6% oxygen. Since for every unit of oxygen two units of CO are produced, the raceway gas amounts to 6500x(1+0.256)=8164 m³ STP. This gas has a higher temperature (decreasing from some 2200°C to 125°C top gas temperature), the furnace is operated at a higher pressure (say 4.8 bar, absolute at the tuyeres and 3 bar, absolute at the top) and extra gas is formed by the direct reduction reaction (see exercise 2.3.5). If all these effects are neglected, the exercise is straightforward: Suppose the void fraction in the burden is 30%, then the open volume in the furnace is (3800–100)x 0.30 = 1100 m³ STP, through which 8164 m³ STP gas is blown per minute. So the residence time of the gas is (1100/8164)x60 = 8 seconds.

It is possible to make the corrections mentioned above. Take an average temperature of the gas of 900°C and an average pressure of 4 bar, and then the effects are:
– Increase in residence time owing to higher pressure: 4/1 = 4 times longer.
– Decrease in residence time owing to higher temperature 273/(273+900)= 0.23 times shorter.
– Decrease in residence time due to extra gas from direct reduction is 8164/9987 = 0.82 times shorter.
– In total, the residence time is shorter by a factor of 0.75 (4x0.23x0.82), so the corrected residence time is 8x0.75 = 6 seconds.

2.3.6 If you get so much top gas, is there a strong wind in the furnace?

No, at the tuyeres there are high wind velocities (over 200 m/sec), but top gas volume is about 9970 m³ STP/min. Over the diameter of the throat, at a gas temperature of 120°C and a top pressure of 2 bar, top gas velocity is 1,0 m/sec: on the Beaufort scale this corresponds to a wind velocity of 1. Through the voids the velocity is about 3 m/s. Note, that in the centre the velocity can be much higher, so that even fluidisation limits can be reached (See 7.4).

III *The Ore Burden: Sinter, Pellets, Lump Ore*

3.1 Introduction

In the early days of commercial ironmaking, blast furnaces were often located close to ore mines. In those days, blast furnaces were using local ore and charcoal, later replaced by coke. In the most industrial areas of the time, the 19th century, many blast furnaces were operating in Germany, Great Britain and the United States. After the application of the steam engine for ships and transportation, the centre of industrial activity moved from the ore bodies to the major rivers, such as the river Rhine, and later from the rivers to the coastal ports with deep sea harbours. This trend, supported by seaborne trade of higher quality ores may appear clear at present, but has only a recent history. In 1960 there were sixty operating blast furnaces in Belgium and Luxembourg. In 2008, only four are operating, of which two have the favourable coastal location.

The trend towards fewer but larger furnaces has made the option for a rich iron burden a more attractive one. A rich iron burden translates into a high Fe content and as fine ores are too impermeable to gas, the choice is narrowed down to sinter, pellets and lump ores. Sinter and pellets are both formed by agglomerating iron ore fines from the ore mines and have normally undergone an enrichment process, which is not described here. The quality demands for the blast furnace burden are discussed and the extent to which sinter, pellets and lump ore meet these demands is described.

A good blast furnace burden consists, for the major part, of sinter and/or pellets (Figure 3.1, next page). Sinter burdens are prevalent in Europe and Asia, while pellet burdens are used more commonly in North America and Scandinavia. Many companies use sinter as well as pellets, although the ratios vary widely.

| Sinter | Pellets | Lump |
| 90 % < 25 mm | 11 mm (± 2 mm) | 6–25 mm |

Figure 3.1   Burden materials

Lump ores are becoming increasingly scarce and generally have poorer properties for the blast furnace burden. For this reason it is used mainly as a cheap replacement for pellets. For high productivity low coke rate blast furnace operation the maximum lump ore rate is in the range of 10 to 15 %. The achievable rate depends on lump ore quality and the successful use of higher percentages is known. The present chapter deals with ore burden quality.

3.2 Iron ore

Iron is the fourth most abundant element in the earth crust, making up approximately 5 % of the total. However, mining of iron (as oxide) is only economically viable where substantial concentration has occurred, and only then can it be referred to as iron ore. More than 3 billion years ago, through the generation of Banded Iron Formation the first concentration occurred. The conventional concept is that in those days the banded iron layers were formed in sea water as the result of an increase in oxygen to form insoluble iron oxides which precipitated out, alternating with mud, which later formed cherts and silicate layers.

Figure 3.2   Banded Iron Formation (National Museum of Mineralogy and Geology, Dresden, picture by André Karwath, file from the Wikimedia Commons)

Subsequently, leaching out of the cherts and silicates resulted in a concentration of the iron oxide and through further geological processes such as (de) hydration, inversion leaching, deformation and sedimentation a wide variety of iron ore deposits have been created all around the world. These total over 300 billion tonnes at an average Fe content of 47 %.

A minor fraction of these deposits are currently commercially mined as iron ore with Fe contents ranging from as low as 30 % up to 64 % (pure iron oxide as haematite contains 70 % Fe). As mentioned before, an efficient blast furnace process requires a rich Fe burden, preferably in excess of 58 % Fe. This material also needs to be within certain size fractions suitable for; pelletizing (indicative <150 μm); sintering (indicative between 150 μm and 6 mm); or as lump ore (indicative between 8 mm and 40 mm) for direct charge. Consequently, the majority of the mined iron ore requires beneficiation and processing prior to becoming a usable material for the blast furnace. This comprises, as a minimum, crushing and screening but most of the time also upgrading and sometimes processing, such as pelletizing at the mine site.

A vast amount of equipment has been developed to economically upgrade the iron ore to a suitable product. These processes will not be described here, but most of them are based on liberating the iron oxide from the gangue minerals and then making use of the differences in density, magnetic properties or surface properties between these minerals to separate them. Sometimes vast amounts of quartz (SiO_2) need to be removed, or minor amounts of impurities (such as phosphorus in the mineral apatite). Depending on the specific requirements, these processes can be easily achieved, or they can be impossible.

These processes result in a wide variety of beneficiated iron ores with varying grades and impurities to be chosen from. Silica content can vary between 0.6 % to above 10 % and phosphorus from below 0.05 % to above 1 %. Similar variations apply for other components such as alumina, lime, magnesium, manganese, titanium and alkalis. With tighter environmental control over the whole process chain, tramp elements at minute levels are starting to play a more dominant role. From sulphur, zinc and copper to mercury, arsenic and vanadium. The importance of these elements greatly depends on the applied process and process conditions, environmental measures and local legislation of where the ores are to be used.

Together with the coal, coke and other plant revert materials, the blast furnace requires a certain burden composition to achieve a balance with respect to all the above elements.

3.3 Quality demands for the blast furnace burden

The demands for the blast furnace burden extend to the chemical composition and the physical durability of the burden materials. The chemical composition must be such that after the reduction and melting processes the correct iron and slag compositions are produced, and this will be determined by the chemical composition of all the materials charged in the furnace. The physical aspects of the quality demands are related to the properties in both the cold and the hot state, and both aspects are discussed in depth in this chapter.

3.3.1 Generation of fines, reducibility, softening and melting

In the shaft zone of the blast furnace the permeability of the burden is determined by the amount of fines (see Figure 3.3). Fines may be defined as the fraction of the material under 5 mm, since the burden components have a general range of 5 to 25 mm. If there are too many fines, the void fraction used for the transport of the reduction gas will reduce and will affect the bulk gas flow through the burden (Hartig et al, 2000). There are two sources for fines, those that are directly charged into the furnace, and those that are generated in the shaft by the process.

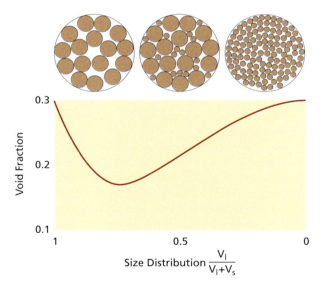

Figure 3.3 *Permeability for gas flow depends on void fraction, which depends on the ratio of smaller and larger particles. Example of two types of spherical particles, large (V_l) and small (V_s). The x–axis gives the fraction of the large particles: $V_l/(V_l+V_s)$.*

During the first reduction step from haematite to magnetite the structure of the burden materials weakens and fines are generated. Sinter and lump ore are especially prone to this effect, known as reduction–disintegration. The

reduction–disintegration depends on the strength of the bonds between the particles of ore fines in sinter and lump ore. Generally speaking, the reduction disintegration is dependent on:
– The FeO percentage in the sinter. The more magnetite (Fe_3O_4, which corresponds with $FeO.Fe_2O_3$) is present, the stronger the sinter. The reduction disintegration takes place at low temperature caused by the change in crystal structure from haematite to magnetite. The FeO percentage in the sinter can be increased by cooling sinter with air that is poor in oxygen. In an operating plant, the FeO in the sinter can be increased by adding more fuel (coke breeze) to the sinter blend.
– The chemical composition of the gangue: basicity, Al_2O_3 and MgO content play an important role.
– The heating and reduction rate of the sinter. The slower the progress of heating and reduction, the higher the reduction disintegration of sinter and lump ore.
– The amount of hydrogen in the reducing gas. More hydrogen in the reducing gas leads to lower reduction disintegration.

A major requirement for the blast furnace ore burden is to limit the quantity of fines within the furnace to as low as possible. This can be achieved by;
– Proper screening of burden materials before charging. Screens with around 5 mm holes are normal operational practice.
– Good reduction–disintegration properties.

During charging, fines in the burden material tend to concentrate at the point of impact on the burden surface. The level of reduction–disintegration increases in areas where the material is heated and reduced slowly. A charged ring of burden with a high concentration of fines will impede gas flow, experience the slower warm–up and so result in a higher level of reduction–disintegration.

The reducibility of the burden is controlled by the contact between gas and the burden particles as a whole, as well as the gas diffusion into the particles. Whether or not good reduction is obtained in the blast furnace is governed by the layer structure of the burden and the permeability of the layers, which determines the blast furnace internal gas flow. This is discussed in depth in the later blast furnace chapters. The reducibility of the burden components will be of less importance if the gas flow within the furnace does not allow sufficient contact for the reactions to take place.

As soon as burden material starts softening and melting, the permeability for gas is greatly reduced. Therefore, the burden materials should start melting at relatively high temperatures and the interval between softening and melting should be as short as possible, so that they do not impede gas flow while they are still high up the stack. Melting properties of burden materials are determined by the slag composition. Melting of acid pellets and lump ore starts at temperatures of 1050 to 1100 °C, while fluxed pellets and basic sinter generally starts melting at higher temperatures. See also section 8.7 on how iron ore melts.

3.3.2 Ore burden quality tests

Ore burden material is characterised by the following.
- Chemical composition.
- Size distribution, which is important for the permeability of ore burden layers in the furnace.
- Metallurgical properties with respect to:
- Cold strength, which is used to characterise the degradation of ore burden materials during transport and handling.
- Reduction–disintegration, which characterises the effect of the first reduction step and is relevant in the stack zone of the furnace.
- Softening and melting properties, which are important for the formation of the cohesive and melting zone in the furnace.

It is important for permeability to have a narrow size range and have minimal fines (less than 3% below 5 mm, after screening in the stockhouse). Measurement of the percentage of fines after screening in the stockhouse can give an indication whether or not excessive fines are charged into the furnace. Material from the stockyard will have varying levels of fines and moisture and thus screening efficiency will be affected accordingly.

A short description of tests used for characterisation of materials is given below with the objective being to understand the terminology.

Principle of tumble test: Sample is tumbled at fixed number of rotations. Size distribution determined after tumbling. Weight percentages over or below certain screen sizes are used as a quality parameter.

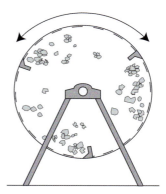

Figure 3.4   *Principle of tumbler test*

	What is measured?	Results	Optimum Range		Reference
			Sinter	Pellets	
Mean Size	Size distribution	Average size, mm % 6.3–16 mm % < 0.5 mm	< 2 %	> 95 % < 2 %	ISO 4701
Cold Strength	Size distribution after tumbling Compression	% > 6.3 mm % < 0.5 mm daN/p	> 70 %	> 95 % < 5 % > 150	ISO 3271 ISO 4700
Strength after reduction LTD (Low Temp. Disintegration)	Size distribution after reduction and tumbling	% > 6.3 mm % < 3.15 mm % < 0.5 mm	< 20 %	> 80 % < 10 %	ISO 4696
Reducibility	Weight decrease during reduction	%/min	> 0.7 %	> 0.5 %	ISO 4695

Table 3.1   Characterisation of ore burden

3.3.2.1 Tests for cold strength

Cold strength is mostly characterised by a tumbler test. For this test an amount of material is tumbled in a rotating drum for a specified time interval. Afterwards the amount of fines are measured. The size distribution after tumbling is determined and used as a quality indicator (Figure 3.4).
For pellets the force needed to crack the pellets, referred to as the Cold Compression Strength, is determined. Although not representative for the blast furnace process, it is a fast and easy test to carry out. The percentage weak pellets give an indication on the quality of induration.

3.3.2.2 Tests for reduction–disintegration

The reduction–disintegration tests are carried out by heating a sample of the burden to at least 500 °C and reducing the sample with gas containing CO (and sometimes H_2). After the test the sample is cooled, tumbled and the amount of fines is measured. The quoted result is the percentage of particles smaller than 3.15 mm.

The HOSIM test (blast furnace simulation test) is a test where the sample is reduced to the endpoint of gas–reduction in a furnace. After the test the sample is then tumbled. The results are the reducibility defined by the time required to reduce the sample to the endpoint of gas reduction, and the reduction–disintegration is represented by the percentage of fines (under 3.15 mm) after tumbling. Although both test are relevant for the upper part of the blast furnace process, the first is excellent to have a daily control on burden quality, but the more advanced HOSIM tests gives a more realistic description of the effects in the blast furnace.

3.4 Sinter

3.4.1 Description

Sinter is made in three different types: acid sinter, fluxed and super–fluxed sinter. Fluxed sinter is the most common type. Since sinter properties vary considerably with the blend type and chemical composition, only some qualitative remarks can be made.

The sinter quality is defined by:
– Size distribution: sinter mean size ranges from 15–25 mm as measured after the sinter plant. The more basic the sinter, the smaller the average size. Sinter degrades during transport and handling so sinter has to be re–screened at the blast furnaces to remove the generated fines. Sinter from stockyard may have different properties from freshly produced sinter directly from the sinter plant. If stock sinter must be used in the blast furnace, it should be charged in a controlled fashion, and diluted with as much fresh sinter as is possible, such as by using a dedicated bin in the stockhouse to stock sinter.
– Cold strength: normally measured with a tumble test. The more energy that is used in the sinter process, the stronger the sinter. The cold strength influences the sinter plant productivity because a low cold strength results in a high fines recycle rate.
– Reduction–disintegration properties. The reduction from haematite to magnetite generates internal stresses within a sinter particle. The stronger the sinter, the better the resistance to these stresses. The reduction–disintegration properties improve with denser sinter structure, i.e. when the sinter is made with more coke breeze. As a consequence of the higher coke breeze usage the FeO content of the sinter will increase. From experimental correlations it is well known, that for a given sinter type, reduction–disintegration improves with FeO content. However, reducibility properties are adversely affected.

The softening and melting of sinter in the blast furnace is determined by the chemical composition, that is the local chemical composition. The three most critical components are the basicity; the presence of remaining FeO; and SiO_2. The latter two function as components that lower the melting temperature. At temperatures of 1200 to 1250 °C sinter starts softening and melting. Very basic parts ($CaO/SiO_2 > 2$) melt at higher temperatures, but will still have melting temperatures around 1300 °C in the presence of sufficient FeO. If, due to further reduction FeO is lowered, then melting temperatures exceeding 1500°C can be observed. However, final melting in a blast furnace differs from melting of "pure" burden materials, since strong interactions between different burden components (super–fluxed sinter and acid pellets) are known to occur.

3.4.2 Background of sinter properties

Sinter is a very heterogeneous type of material. Research of various types of sinter in a cooled furnace has demonstrated that various phases are present simultaneously, see Figure 3.7. The most important phases present are:
- Primary and secondary magnetite (Fe_3O_4). Secondary magnetite is formed during sintering in the high temperature, reducing areas at the sinter strand, those being areas in close proximity to coke.
- Primary and secondary haematite (Fe_2O_3). Secondary haematite is formed on the sinter strand during the cooling down of the sinter in the presence of air (oxygen).
- Calcium ferrites are structures formed from burnt lime (CaO) and iron oxides.

It is clear from Figure 3.5, that at increasing basicity an increased fraction of calcium ferrites can be found. This has major consequences, for the sintering process as well as for the use of sinter in the blast furnace.

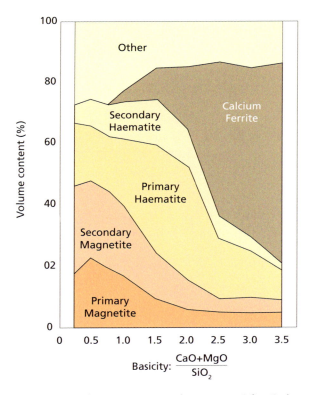

Figure 3.5   Phase composition of sinter types (after Grebe et al, 1980)

Firstly, let us consider the liquidus temperatures of sinter–type materials. The acid sinter has much higher liquidus temperature than basic sinter. This is due to the fact that calcium ferrite type structures have liquidus temperatures as low as 1200 °C (Figure 3.6), while the acid sinter have liquidus temperatures well above 1400 °C. It means also, that sintering of fluxed or superfluxed sinter can be accomplished at lower temperatures than sintering of a more acid sinter blend. Because of this, acid sinter is generally coarser and has a higher cold strength than basic sinter.

The reason why high basicity sinter is formed at much lower temperature than acid sinter is illustrated in Figure 3.6, where a diagram of FeOn with CaO is shown. FeOn means a combination of Fe and FeO and Fe_2O_3. During sintering, coke breeze is burnt and locally, a reducing atmosphere exists, which reduces Fe_2O_3 to FeO. On specific location, the chemical composition is such, that melts with very low melting temperatures can be formed. In Figure 3.6 it is shown, that at weight percentages of over 15 % CaO, melting temperatures as low as 1070 °C can be found. If less CaO is present, the melting temperature is much higher, i.e. 1370 °C. This is where acid sinter is made.

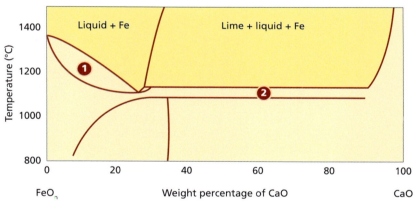

Figure 3.6   Formation of liquid phases in a mixture of Lime (CaO) and iron oxide (FeO_n) – FeO_n represents a mixture of iron (Fe), wüstite (FeO) and haematite (Fe_2O_3) (after Allen & Snow, Journal of the American Ceramic Society volume 38 (1955) Number 8, page 264)

Next, we consider the reduction–disintegration properties of the sinter. The driving force of low temperature reduction–disintegration of sinter is the changeover of the crystal structure from haematite to magnetite, which causes internal stress in the iron oxide crystal structure. So, reduction–disintegration of sinter is related to the fraction of haematite in the sinter. As shown in Figure 3.5, there is primary and secondary haematite in the sinter. Particularly the latter causes reduction–disintegration, since it is more easily reduced in the upper part of the furnace than primary haematite (see Figure 3.7).

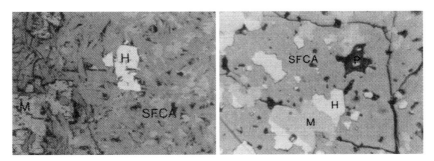

Figure 3.7   Cracking of calcium ferrites (SFCA) due to reduction of primary (left) and secondary (right) haematite (H) into magnetite (M). Pores appear black. (Chaigneau, 1994)

The higher the secondary haematite percentage in the sinter, the more the sinter is prone to reduction–disintegration effects. This can also be said in reverse, that is, there is a strong relationship between the FeO content of the sinter and the reduction–disintegration. The higher the FeO content, the less reduction disintegration will take place. The FeO content of sinter can be increased by adding more fuel to the sinter blend, which is normally done in the form of coke breeze. However, the precise relationship between the FeO content of the sinter and the sinter quality depends on the ore blend used and is plant–specific. The reduction–disintegration properties depend on the type of FeO present in the crystal structure. To illustrate this by example; a high fraction of magnetite in the sinter blend will give sinter with a high (primary) magnetite fraction. Moreover, in the presence of sufficient SiO_2 fayalite structures ($2FeO.SiO_2$) can be formed. These structures are chemically very stable and can only be reduced at high temperatures by direct reduction reactions (see section 8.2.1). Alternatively, in the presence of MgO, spinel structures containing large amounts of FeO can be formed. These spinel structures are relatively easy to reduce. Finally, sinter that has been formed at high temperatures (acid sinter), will contain glass–like structures where the FeO is relatively difficult to reduce.

It is possible to suppress the formation of secondary haematite by cooling the sinter with air–gas mix with a reduced oxygen percentage (12 to 14%). This results in a relatively high FeO content of the sinter, because less secondary haematite is formed. This has a major benefit for the reduction–disintegration properties of this type of sinter. In addition, the calorific value of the blast furnace top gas increases, as less oxygen has been removed from the ore burden, giving an economic advantage.

During the sintering process there is a major difference between the use of CaO and MgO as fluxes. Both materials are normally added as the carbonate, using limestone as $CaCO_3$ or dolomite as $CaCO_3.MgCO_3$. The carbonates are decomposed on the sinter strand, requiring a large energy input. However, the melts containing substantial amounts of CaO have low liquidus temperatures, such as 1100 °C for mixtures of 20 to 27 % CaO and iron oxides. For the melts containing MgO, the spinel structures mentioned above, the melting temperatures are much higher. Therefore, it is easier to form slag–bonds in the sinter using CaO than with MgO. And generally, making sinter with CaO can be done at lower temperature. But sinter with high MgO is more resistant against reduction–disintegration. MgO content can be increased by adding olivine of serpentine to the sinter blend.

For the final result of the produced sinter, it is important to note that the sinter blend prior to sintering is far from homogeneous. It contains various types of material and locally there are widely varying compositions and sizes present. Ore particles can be as large as 5 mm, coke breeze up to 3 mm and limestone and dolomite up to 2.5 mm. All types of chemical compositions are present on the micro–scale, where the sintering takes place. Types of materials used, size distribution of the various materials, the blending of the sinter mix, the amount of slag–bonds forming materials in the blend as well as the amount of fuel used for the sintering all have specific disadvantages for good sinter quality. This makes optimisation of sinter–quality a plant–specific technological challenge.

In the above sections the importance of reduction–disintegration of sinter is stressed. The lower the reduction–disintegration, the poorer the reducibility of the sinter. Needle–like structures of calcium ferrites have a relatively open structure and are easily accessible for reduction gas in the blast furnace. In cold conditions the sinter is strong (i.e. good tumbler test results), the degradation during transportation is also good, but the relatively fast reduction in the blast furnace makes the sinter very prone to reduction–disintegration. More solid structures in the sinter have better properties in this respect. Reduction–disintegration leads to poorer permeability of the ore layers in the furnace and impedes proper further reduction of the iron oxides in the blast furnace.

3.5 Pellets

3.5.1 Pellet quality

With correct chemical composition and induration, pellets can easily be transported from mine to blast furnace, can be stocked and remain generally intact in the blast furnace. Therefore, when judging pellets the main issues are:
– Cold strength, measured as compression strength and the fines generated through tumbling. Low figures indicates bad or lean firing.
– The reduction–disintegration properties. These properties are less of a concern with pellets than with sinter and lump ore.
– The swelling properties. With incorrect slag composition pellets tend to have extreme swelling properties. Since the phenomenon is well known, it normally does not happen with commercially available pellets.
– The softening and melting. Pellets tend to melt at lower temperatures than

fluxed sinter.

Alongside proper induration, the slag volume and composition and the bonding forces mainly determine the quality of pellets. The three main pellet types are:
- Acid pellets
- Basic pellets
- Olivine doped pellets

Typical properties of the three types of pellets are shown in Table 3.2.

Pellet Type	Compression	Reducibility	Swelling
Acid	++	-	+/-
Basic	+	+	+
Olivine	+	+	+

Pellet Type	Fe %	SiO$_2$ %	B	MgO %	Compression (kg/pellet)
Acid	65-67	2-5	<0.5	0.1-0.6	>270
Basic	63-66	1.5-4	0.8-1.1	0.1-1.5	>240
Olivine	64-67	2-4	<0.5	1.3-1.8	>180

Table 3.2   Overview pellet properties

Acid pellets are strong, but have moderate metallurgical properties. They have good compression strength (over 250 kg/pellet), but relatively poor reducibility. In addition, acid pellets are very sensitive to the CaO content with respect to swelling. At CaO/SiO$_2$ > 0.25 some pellets have a strong tendency to swell, which might jeopardize proper blast furnace operation.

Basic and fluxed pellets have good metallurgical properties for blast furnace operation. By adding limestone to the pellet blend, the energy requirement of the firing/induration increases because of the decarbonisation reaction. For this reason production capacity of a pellet plant can sometimes be 10 to 15% lower when producing basic pellets compared with acid.

Olivine pellets contain MgO in place of CaO, which is added to the blend as olivine or serpentine. The pellets are somewhat weaker when tested for cold compression strength.

3.5.1.1 Cold compression strength

The difference in compression strength might seem large. However, in the blast furnace the pellets are reduced and the difference diminishes during reduction. After the first reduction step to Fe$_3$O$_4$, the cold compression strength drops to 45–50 kg for acid pellets and to 35–45 kg for olivine pellets. Therefore, a little lower average compression strength has no drawback for the blast furnace process as long as it is not caused by an increased percentage of very weak pellets (< 60 kg/pellet). Especially that fraction is a good indicator for the pelletizing process: the more pellets that collapse at low compression strength, the poorer the pellets have been fired. Therefore, pellet quality can be influenced by the production rate: the slower the grate is moving the stronger the firing can be, so the induration period increases and the pellets become stronger.

3.5.1.2 Swelling

As mentioned above, pellets, in contrast to sinter and lump ore, can have the tendency to swell during reduction. Generally a volume increase of over 20 %, measured according to ISO 4698, is seen as critical. The effect, however, depends on the percentage of pellets used in the burden. Swelling occurs during the transformation of wustite into iron, but like any transformation, this is a balance between iron nucleation and growth of these nuclei. During swelling, limited nucleation occurs and these nuclei grow like needles causing a volume increase which is seen as swelling, see figures 3.8 and 3.9. These needles are difficult to observe. Under certain conditions, for example in the presence of alkalis in the blast furnace, the swelling can become excessive and a cauliflower–structure develops. This coincides with a low compression strength of this structure, with the opportunity to generate fines.

Figure 3.8 *Limited swelling accompanied by the formation of an iron shell.*

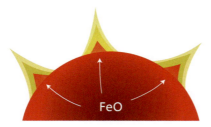

Figure 3.9 *Limited iron nucleation followed by strong needle growth of the nuclei with as a result excessive swelling of the pellet.*

Main factors influencing pellet swelling are basicity and gangue content. Figure 3.10 shows how swelling depends on pellet basicity. Pellets with a basicity between 0.2 and 0.7 are more prone to swelling.

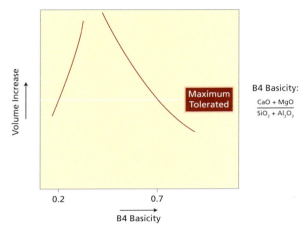

Figure 3.10   *Graph showing volume increase effect of pellet swelling with increasing basicity of the pellet.*

Swelling is mitigated by proper induration. In the blast furnace local process conditions like temperature and gas composition greatly influence the swelling behaviour. At higher reduction degrees swollen pellets shrink. As the phenomenon of swelling is well known, it is normally under control with commercially available pellets, but always requires a check because it could have a severe impact on the regularity of the blast furnace process.

Each process demands its specific optimum pellet quality, but a summary of acceptable ranges is given in Table 3.3 bearing in mind the earlier mentioned differences between the pellet types.

	What is measured?	Results	Acceptable Range	Reference
Mean Size	Size distribution	% 6.3–16 mm % < 6.3 mm	> 95 % < 2 %	ISO 4701
Cold Strength	Compression Strength Tumbling Strength and Abrasion	Average kg/p % < 60 kg/p % > 6.3 mm % < 0.5 mm	> 150 kg/p < 5 % > 90 % < 5 %	ISO 4700 ISO 3271
LTB (Low Temp. Breakdown)	Size distribution after static reduction and tumbling	% > 6.3 mm	> 80 %	ISO 4696
Reducibility	Weight decrease during reduction	%/min $(dR/dt)_{40}$	> 0.5 %/min	ISO 4695

Table 3.3   *Characterisation of pellets*

3.6 Lump ore

Lump ores are natural iron–rich materials, which are used directly from the mines. Because the lump ores are screened out at the mines, the mines generally produce lump ore as well as (sinter) fines. Major lump ore deposits are present in Australia (Pilbara region), South America (Carajas and Iron Ore Quadrangle), and South Africa (Sishen). In many other places limited amounts of lump ores are produced. Lump ores are becoming more and more scarce.

The lump ores are cheaper than pellets. For this reason in many blast furnaces high amounts of lump ore are being considered. The lower cost of the lump ore compared with pellets is offset by the poorer metallurgical properties. Generally speaking, in comparison with pellets, lump ores:
– Show some decrepitation due to evaporating moisture in the upper stack of the furnace
– Generate more fines during transport and handling.
– Have poorer reduction degradation properties and may have poorer reducibility properties.
– Have a lower melting temperature.
– Have greater diversity in physical properties due to being naturally occurring

Lump ore is used in an appropriate size fraction, such as 8–30 mm.
For blast furnace operation at high productivity and high coal injection levels, lump ore is not the preferred burden material. As lump ore is a natural material, properties can differ from type to type. Certain types of lump ores can compete favourably with sinter, and in the case of Siderar blast furnace in Argentina, they have operated successfully with up to 40% in the burden of a Brazilian lump ore at high furnace productivity.

3.7 Interaction of burden components

The results of burden tests on the total burden can differ greatly from results on sinter, pellets and lump ore alone. An example is given in Figure 3.11. A relatively poor quality of lump ore is blended with good sinter. It is shown that the behaviour of the blend is better than expected from the arithmetic mean of the data. Generally speaking, blending of materials dilutes the disadvantages of a certain material. Therefore, the blast furnace burden components have to be properly blended when charged into the furnace.

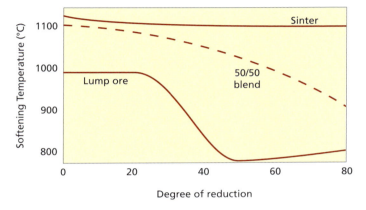

Figure 3.11   *Softening temperature of a 50/50 blend of sinter and lump ore (Example taken from Singh et al, 1984)*

IV *Coke*

4.1 Introduction: function of coke in the blast furnace

Coke is basically a strong, non–melting material which forms lumps based on a structure of carbonaceous material internally glued together (Figure 4.1).

Figure 4.1   Coke

The average size of the coke particles is much larger than that of the ore burden materials and the coke will remain in a solid state throughout the blast furnace process.

For blast furnace ironmaking the most important functions of coke are:
– To provide the structure through which gas can ascend and be distributed through the burden. Coke is a solid and permeable material up to very high temperatures (> 2000 °C), which is of particular importance in the hearth and melting and softening zone. Below the melting zone coke is the only solid material, so the total weight of the blast furnace content is supported by the coke structure. The coke bed has to be permeable, so that slag and iron can flow downward to accumulate in the hearth and flow to the tap hole.
– To generate heat to melt the burden
– To generate reducing gases
– To provide the carbon for carburization of the hot metal
– To act as a filter for soot and dust.

The permanent efforts aimed at reducing the costs of iron making have lead to an increasing portion of substitute reduction materials for coke, which has mainly been coal injected through the tuyeres. Nowadays, blast furnaces with total coal injection rates in excess of 200 kg/tHM are operated with coke consumptions of less than 300 kg/tHM. At these high coal injection rates, coke is subjected to more rigorous conditions in the blast furnace. Dissection of furnaces taken out of operation and probing and sampling through the tuyeres of furnaces in operation have allowed the assessment of the extent of coke degradation in the furnace. Coke degradation is controlled by the properties of feed coke, i.e. mechanical stabilization, resistance to chemical attack (solution loss, alkalis, and graphitization) and by the blast furnace operating conditions. At high coal injection rates the amount of coke present in the furnace decreases and the remaining coke is subjected to more vigorous mechanical and chemical conditions: increased mechanical load as the ore/coke ratio becomes higher; increased residence time at high temperatures; increased solution loss reaction (CO_2, liquid oxides); and alkali attack. More severe coke degradation during its descent from the furnace stock line into the hearth can therefore be expected at high coal rates.

However, high coal injection rates can also affect the direct reduction reactions.
1. Coal injection increases hydrogen content and at elevated temperatures (800–1100 °C), hydrogen is a very effective agent in gas reduction of iron oxides.
2. The unburnt soot remaining after the raceway is more reactive than coke and used for direct reduction in preference of coke.
3. The alkali cycle is reduced as a consequence of the elimination of alkali through the hot furnace centre.

Therefore, at high coal injection rates the attack of coke by direct reduction reactions may also decrease. This is beneficial for coke integrity in the lower part of the furnace.

In this chapter we will discuss coke quality parameters, test methods, degradation processes of the coke in the blast furnace, and finally the range of coke qualities targeted by blast furnaces who are or are aiming to operate at the highest production levels, so are more demanding in terms of coke quality.

4.2 Coal blends for coke making

The coal selected to make coke is the most important variable that controls the coke properties. The rank and type of coal selected impacts on coke strength while coal chemistry largely determines coke chemistry. In general, bituminous coals are selected for blending to make blast furnace coke of high strength with acceptable reactivity and at competitive cost. For the conventional recovery coking process the blend must contract sufficiently for easy removal from the oven and pressure must be acceptable. For the heat–recovery process type these constraints are not valid, which leads to an increase of usable coal types in this type of process. Table 4.1 shows the typical chemical composition of coke that may be considered to be of good quality.

Typical Coke Analysis		% (db)
Coke Analysis	Fixed Carbon	87–92
	Nitrogen	1.2–1.5
	Ash	8–11
	Sulphur	0.6–0.8
	Volatile Matter (for well carbonised coke)	0.2–0.5
Ash Analysis	Silica SiO_2	52.0
	Alumina Al_2O_3	31.0
	Iron Fe	7.0
	Lime CaO	2.5
	Potassium K_2O	1.8
	Magnesia MgO	1.2
	Sodium Na_2O	0.7
	Phosphorous P	0.3
	Manganese Mn	0.1
	Zinc Zn	< 0.02

Table 4.1   Coke chemistry for a typically acceptable coke quality grade

Ash directly replaces carbon. The increased amount of slag requires energy to melt and more fluxes to provide a liquid slag. Ash, sulphur, phosphorous, alkalis and zinc can be best controlled by careful selection of all coal, coke and burden materials. The financial repercussions of ash, sulphur and phosphorous may be assessed by value–in–use calculations for PCI–coal, coking coal blends and burden materials. Alkalis and zinc should remain below certain threshold levels (Section 6.2).

4.3 Coke quality concept

Now the question is: how to characterize coke quality; how to define and measure the coke properties. In other words, how to establish a target for coke manufacturing based on determined coke properties in line with the needs of the blast furnace process. From the above discussion, the following parameters should be considered to limit the coke degradation and maintain suitable coke behaviour in the blast furnace, especially at high coal injection rates. Qualitatively the coke should:
– Be made up of large, stabilized particles within a narrow size distribution band
– Have a high resistance against volume breakage
– Have a high resistance against abrasion
– Have a high resistance against chemical attack (CO_2, Alkali)
– Have a high residual strength after chemical attack
– Have sufficient carburization properties (the dissolution of carbon in hot metal).

4.3.1 Coke degradation mechanisms in the blast furnace

The basic concepts of coke degradation in the blast furnace, according to the interconnected thermal, physical, and chemical conditions coke is subjected to in the furnace are described in Figure 4.2.

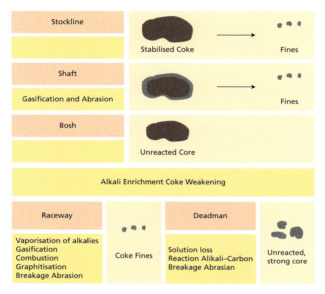

Figure 4.2   Basic concepts of coke degradation in a blast furnace

At the stockline, the coke is generally well stabilized. The effect of gasification on strength is controlled by the mechanisms of the heterogeneous reaction. In general, diffusion is the limiting step and the reaction is located at the surface of the lumps, the core remaining quite unaffected. As gasification and abrasion proceed simultaneously, a peeling of coke particles occurs (3 – 5 mm size reduction), leaving an exposed unreacted core and fines.

Beyond gasification, coke reacts with alkali vapours when passing through the alkali circulating zone and the structure is penetrated by alkalis. This reaction reduces the strength of the coke, making it more susceptible to size reduction by breakage from mechanical action. Coke that has been already weakened arriving in the high temperature zone of raceway loses its alkalis by gasification. High temperature, mechanical action and graphitization bring about severe degradation, decrease of size and formation of fines.

The coke travelling to the dead man is exposed to moderate temperatures, high alkalis during long periods of time along with additional reactions (reduction of slag, carburization) that mostly effect the surface of the coke lumps. Dead man coke, sampled by core drilling corresponds more or less to the unreacted core of the initial lumps and it is not surprising that it exhibits similar strength to the coke that is charged at the top.

4.3.2 Degradation of coke during its descent in the blast furnace

To discuss the phenomena leading to coke degradation during descent in the blast furnace we make use of Figure 4.3 representing the different zones of the process, the relevant process conditions and the development of the coke size under these conditions.

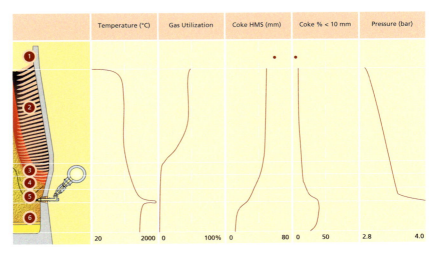

Figure 4.3   Development of coke size under the conditions that are present in the blast furnace throughout the journey from the top to the bottom of the furnace.

1. Charging zone: Due to the fall of the coke onto the stockline some breakage and abrasion will occur during charging.
2. Granular zone: In this region the coke and ore remain as discrete particles within their separate layers. Drying occurs and recirculating elements such as zinc, sulphur and alkalis deposit on the burden materials as they descend to the bottom of the granular zone. From a temperature of 900 °C coke starts to oxidize with CO_2, continuing to do so as the temperature increases to over 1000 °C. In this zone coke degradation (mostly abrasion) occurs due to mechanical load and mild gasification.
3. Cohesive zone: This zone starts where ore agglomerates begin to soften and deform, creating a mass of agglomerate particles sticking together. This mass is barely permeable and the rising gas can only pass through the remaining coke layers. Coke gasification with CO_2 becomes significant due to increased reaction rates at the higher temperature level (1000 – 1300 °C). The contact between the softened or molten materials and the coke lumps becomes more intensive, leading to increased mechanical wear on the outer surface of the coke particle. The residence time within the cohesive zone is rather short (30 to 60 minutes) depending on productivity and softening properties of the agglomerates.

4. Active Coke or Dripping zone: This is a packed bed of coke through which liquid iron and slag percolate towards the furnace hearth. The coke particles play an active role in further reducing the remaining iron oxides and increasing the carbon content of the iron through dissolution of carbon from the coke into the iron. The bulk of the coke arriving in this zone (also referred to as bosh coke) flows towards the raceway region. The remaining part will move into the dead man. The residence time estimates varies from 4 to 12 hours. The temperature increases gradually from 1200 to 1500 °C.
5. Raceway: Hot blast containing oxygen is introduced through the tuyeres. The kinetic energy of the blast creates a raceway (cavity) in front of each tuyere. Coke particles circulate at very high velocity in this semi–void region while being gasified together with injectants such as coal, oil and natural gas. A part of the coke and injected reductants is not burnt completely. Soot is produced during injection of coal and natural gas. Soot and dust are transported upwards by the gas stream. They cover coke particles and react later following solution loss reaction. They decrease the reactivity of coke and cause an increase in apparent viscosity of liquid phases. The temperature increases rapidly to over 2000 °C due to the exothermic oxidation of coke and injectants. Coke and injectant fines that are generated in the raceway either completely gasify or get blown out of the raceway into the coke bed. Coke and coal fines may accumulate directly behind the raceway, forming an almost impermeable zone called the bird's nest. Observations of the raceway were made in blast furnaces in operation by inserting an endoscope through a tuyere. These observations showed that in this zone the coke is subjected to very severe conditions.
6. The Hearth: Since the rate of coke consumption is the highest in the ring of the raceway, an almost stagnant zone (not directly feeding the raceway) develops in the furnace centre. This zone is called the dead–man, and is thought to have a conical shape and a relatively dense skin structure. Molten iron and slag accumulates throughout the structure before being tapped through the tapholes. Tracer experiments in a German furnace gave values in the range of 10 to 14 days, but in literature also residence times of 60 days are mentioned for the deadman coke.

The above mentioned processes are summarized in Table 4.2.

Blast Furnace Zone	Function of Coke	Coke Degradation Mechanism	Coke Requirements
Charging Zone		– Impact Stress – Abrasion	– Size Distribution – Resistance to Breakage – Abrasion Resistance
Granular Zone	– Gas permeability	– Alkali Deposition – Mechanical Stress – Abrasion	– Size & Stability – Mechanical Strength – Abrasion Resistance
Cohesive Zone	– Burden support – Gas permeability – Iron and slag drainage	– Gasification by CO_2 – Abrasion	– Size Ditribution – Low Reactivity to CO_2 – High Strength after Abrasion
Active Zone	– Burden support – Gas permeability – Iron and slag drainage	– Gasification by CO_2 – Abrasion – Alkali attack and ash reactions	– Size Ditribution – Low Reactivity to CO_2 – Abrasion Resistance
Raceway Zone	– Generation of CO	– Combustion – Thermal Shock – Graphitisation – Impact Stress and Abrasion	– Strength against Thermal Shock and Mechanical Stress – Abrasion Resistance
Hearth Zone	– Burden support – Iron and slag drainage – Carburisation of iron	– Graphitisation – Dissolution into hot metal – Mechanical Stress	– Size Distribution – Mechanical Strength – Abrasion Resistance – Carbon Solution

Table 4.2   Coke functions, degradation mechamisms and requirements

4.4 Coke size distribution

The shape of the coke particles and the size distribution of the particles are the decisive factors for the permeability of the coke bed, for ascending gas as well as for the descending liquids. Research has shown that the harmonic mean size (HMS), of the coke mass gives the highest correlation with the resistance to flow of gas passing through the coke bed. HMS is the size of uniform size balls with the same total surface as the original coke size mixture.

The lowest flow resistance is obtained when large coke is being used of high uniformity. Fines in particular have a strong decreasing effect on the harmonic mean size and so on the bulk resistance of the coke. Although excellent blast furnace operations are reported with screening at 24 mm (square) there are also plants where screening even at 40 mm is preferred.

Once the coke bulk has been classified by screening and crushing (see also Figure 4.4) the aim is to have a resulting coke with a high mechanical strength under the blast furnace conditions. This is to prevent an excessive formation of coke fines during its descent in the blast furnace.

4.5 Mechanical strength of coke

4.5.1 Coke particle formation and stabilization

During carbonization in a coke oven, fissures in the coke are generated due to stresses that arise from the differential contraction rates in adjacent layers of coke, which are at different temperatures. Typically they are longitudinal, that is perpendicular to the oven walls. Additionally, many transverse fissures are formed during pushing. These fissures determine the size distribution of the product coke by breakage along their lines during subsequent handling. But not all the fissures lead to breakage at this early stage, and a number of them remain in the coke particles. The initial coke distribution is a function of the coal blend and the coking conditions. A significant number of internal fissures remain present and cause further degradation under mechanical loads during transport and charging of the blast furnace. This process of coke degradation is called stabilization. Stabilization lowers the mean size of the coke, but the resulting particles are less prone to further breakage. For blast furnace performance it is not only important to have large, stabilized and narrow size distribution coke charged into the furnace, but it is even more important to have the same qualities present during its descent through the furnace as well. With mechanical handling coke particles will degrade due to breakage and abrasion. Breakage is the degradation of coke by impact due to fissures already present in the coke. Abrasion is the degradation of the surface by relatively low impact processes (rolling and sliding). It is one of the main mechanical processes for decreasing the coke size below the stock line, next to breakage in the race way area. Abrasion causes the formation of fines which may hamper blast furnace permeability.

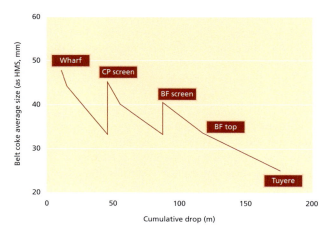

Figure 4.4   *Development of Harmonic Mean Size after mechanical handling in the form of drops between conveyors and screens.*

The resistance to abrasion will deteriorate in the blast furnace, due to reactions such as graphitization, gasification and carburization of the iron. Graphitization results in a more crystalline form of carbon in the coke that is more brittle. In Figure 4.4 the typical development of the HMS of coke from the coke wharf to the tuyeres is presented.

In the presented transport route the coke is screened at 35 mm (square) at the coke plant and at 24 mm (square) at the blast furnace. The increase in HMS of the sample after screening is due to the removal of the undersized coke from the batch.

4.5.2 Coke strength simulation tests

Although it is known that coke degrades more rapidly at high temperatures, there is no test in practical use that is performed at high temperatures. Not only because of the complexity and high costs but also that it has been proven that coke with poor low–temperature strength also exhibit poor strength at high temperatures. Therefore most tests in practical use are done at ambient temperature.

Coke strength is traditionally measured by empirical tumble indices. During mechanical handling coke size degradation takes place by two independent processes, those being breakage into smaller lumps along fissures and cracks still present in the lumps, and abrasion at the coke surface resulting in small particles (< 10 mm). So it is common to measure a 'strength' index related to degradation by volume breakage, for example, I_{40}, M_{40}; and an 'abrasion' index, for example, I_{10}, M_{10}, D^{150}_{15}. These empirical indices cannot be directly related to fundamental coke properties.

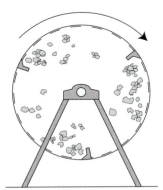

Figure 4.5   Schematic showing the motion of coke in a tumble test

Figure 4.5 shows a schematic representation of particle motion in a tumble drum. As a lifter sweeps around, it picks up a portion of the coke. Some of the coke rolls off the lifter before it reaches the horizontal plane. The coke that is not picked up slips and rolls against the bottom of the drum (a). The coke that is lifted past the horizontal is dropped over a fairly narrow angular range as the lifter approaches the vertical plane (b). This coke impacts with the bottom of the drum. Tests have shown that there is a relationship between the degradation of coke in a drum test and that after a number of drops. This makes it possible to translate the effect on coke size after a number of drops, in metres, into a number of rotations in a drum, and vice versa.

4.6 Overview of international quality parameters

Table 4.3 gives an overview of typical coke quality parameters and their generally accepted levels for a 'good' coke quality. Although not complete, the values given in the table represent coke qualities that have assisted in securing excellent blast furnace results over a long period.

We have to stress, however, that blast furnace operation is very much influenced by coke variability: the gas flow in the furnace can only be held consistent if the layer build–up is consistant and if day to day consistency of the coke is very good. There are, however, no international standards or criteria for day to day consistency.

	What is measured?	Results	Accept. Range	Best	Reference
Mean Size	Size Distribution	AMS mm HMS mm % < 40 mm % < 10 mm	40–60 35–50 < 25 < 2 %		
Cold Strength	Size Distribution after Tumbling	I_{40} % > 40 mm I_{10} % < 10 mm M_{40} % > 40 mm M_{10} % < 10 mm Micum Slope Fissure Free Size DI^{150}_{15}	> 45 < 20 > 80 < 7 0.55–0.7 35–55 84–85	60 16 87 5.5 0.55 85	Irsid Test Micum Test Ext. Micum JIS Test
	Stability at Wharf Stab. at Stockh. Hardness	% > 1" % > 1" % > ¼"	> 58 > 60 > 70		ASTM Test
Strength after reaction	CSR	% > 9.52 mm	> 58	70	Nippon Steel Test
Reactivity	CRI	% weight loss	< 29	22	Nippon Steel Test

Table 4.3   Acceptability range for coke quality parameters
(for tests, see Annex IV)

V Injection of Coal, Oil and Gas

The energy inputs and outputs of the blast furnace are schematically shown in Figure 5.1. The major sources for energy in the furnace are the coke and injectants (coal, gas, oil) and the sensible heat of the hot blast. The major part of the energy is used to drive the change from iron oxides to iron and the other chemical reactions. The remaining energy leaves the furnace as top gas, as sensible heat of iron and slag and as heat losses.

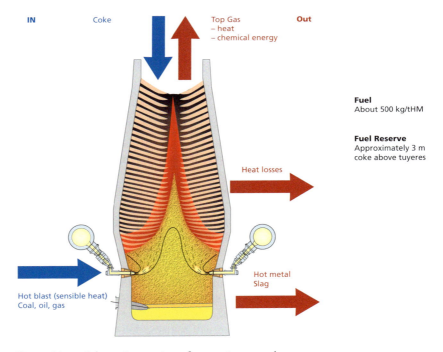

Figure 5.1   Schematic overview of energy inputs and outputs

Use of injection of pulverised (or granular) coal, oil and natural gas can lower the coke rate and thus the cost of hot metal. The auxiliary reductants are mainly coal, oil and natural gas, but tar and other materials can also be used. The precise financial balance depends very much on local situations. Up until the early 1980's oil injection was a commonly used, however the changes in relative prices between coal and oil has resulted in coal becoming the more widely used injectant. Note, that the preparation of coal for injection involves a rather

high investment cost. The pay–back of the investment heavily depends on the hot metal production level. Most major sites have been equipped with coal injection. When coke is scarce and expensive, the feasibility of coal injection for smaller sites increases. The most important arguments for the injection of coal (or natural gas) in a blast furnace are;
– Cost savings by lower coke rates. Cost of coke is substantially higher than that of coal, moreover, the use of an injectant allows higher blast temperatures to be used, which also leads to a lower coke rate.
– Increased productivity from using oxygen enriched blast.
– Decrease of the CO_2 foot print, i.e. the amount of CO_2 produced per ton of steel.

The reason for the apparent versatility of the blast furnace in consuming all types of carbon containing materials is that at the tuyeres the flame temperatures are so high that all injected materials are converted to simple molecules like H_2 and CO and behind the raceway the furnace "does not know" what type of injectant was used.

Coal injection was applied in the blast furnace Amanda of ARMCO (Ashland, Kentucky) in the 1960's. In the early days of coal injection, injection levels of 60–100 kg coal per tonne hot metal were common. Presently, the industrial standard is to reach a coke rate of 300 kg/t with injection levels of 200 kg coal per tonne hot metal (McMaster 2008, Carpenter 2006).

5.1 Coal injection: equipment

The basic design for coal injection installations requires the following functions to be carried out (Figure 5.2):
– Grinding of the coal. Coal has to be ground to very small sizes. Most commonly used is pulverised coal: around 60 % of the coal is under 75 µm. Granular coal is somewhat coarser with sizes up to 1 to 2 mm.
– Drying of the coal. Coal contains substantial amounts of moisture, 8 % to more than 10 %. Since injection of moisture increases the reductant rate, moisture should be removed as much as possible.
– Transportation of the coal through the pipelines. If the coal is too small the pneumatic transport will be hampered. It may result in formation of minor scabs on the walls and also lead to coal leakage from the transportation pipes.
– Injection of the pulverised coal: Coal has to be injected in equal amounts through all the tuyeres. Particularly at low coke rate and high productivity the circumferential symmetry of the injection should be maintained.

There are various suppliers available for pulverised coal injection (PCI) installations, which undertake the functions mentioned above in a specific way. The reliability of the equipment is of utmost importance, since a blast furnace has to be stopped within one hour, if the coal injection stops.

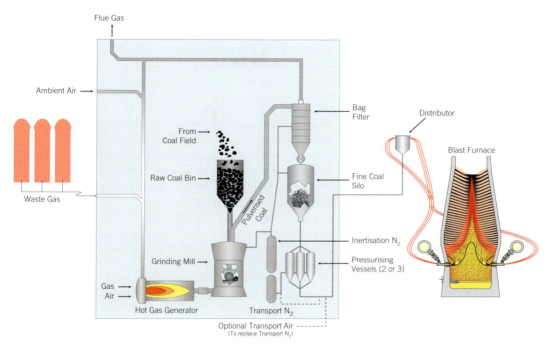

Figure 5.2   Example of PCI installation

5.2 Coal specification for PCI

5.2.1 Coke replacement

Coal types are discriminated according to their volatile matter content. The volatile matter is determined by weighing coal before and after heating for three minutes at 900 °C. Coals that have between 6 and 12 % volatile mater are classified as low volatile, those between 12 and 30 % are mid volatile and anything over 30 % are high volatile. All types of coal have successfully been used. The most important property of the injection coal is the replacement ratio (RR) of coke. The composition and moisture content of the coals determine the amount of coke replaced by a certain type of coal. The replacement ratio of coal can be calculated with a mass and heat balance of the furnace. The chemical composition of the coal (i.e. carbon percentage, hydrogen percentage, ash content), the remaining moisture and the heat required to crack the coal chemical structure (especially the C–H bonds) have to be taken into account. A simplified formula for the replacement ratio (compared with coke with 87.5% carbon) is:

$$RR = 2x\ C\%(coal) + 2.5x\ H\%(coal) - 2x\ moisture\%(coal) - 86 + 0.9x\ ash\%(coal)$$

This formula shows, that the coke replacement depends on carbon and hydrogen content of the coal. Any remaining moisture in the coal consumes energy introduced with the coal. The positive factor of the ash content comes from a correction for heat balance effects.

5.2.2 Coal quality

- Composition: high sulphur and high phosphorous are likely to increase costs in the steel plant. These elements should be evaluated prior to the purchase of a certain type of coal. Young coals (high oxygen content) are known to be more susceptible to self–heating and ignition in atmospheres containing oxygen. This is also an important factor that must be considered with regard to the limitations of the ground coal handling system.
- Volatile matter: high volatile coals are easily gasified in the raceway, but have lower replacement ratio in the process.
- Hardness. The hardness of the coal, characterised by the Hardgrove Grindability Index (HGI) must correspond to the specifications of the grinding equipment. The resulting size of the ground coal is also strongly dependent on this parameter and must correspond to the limits of the coal handling and injection system.
- Moisture content. The moisture content of the raw coal as well as the surface moisture in the ground coal must be considered. Surface moisture in the ground coal will lead to sticking and handling problems.

Potential injection coals can be evaluated on the basis of "value in use", where all effects on cost are taken into account. It is often possible to use blends of two or three types of injection coals, so that unfavourable properties can be diluted.

5.2.3 Coal blending

Most companies use coal blends for injection. Blending allows for (financial) optimization of coal purchases. E.g. a company with a grinding mill for hard coals can use a considerable percentage of softer coals by blending it into hard coals. In doing so, an optimized value can be obtained. Blending dilutes the disadvantages coal types. Every material has disadvantages like high moisture content, sulphur of phosphorous level, a relatively poor replacement ratio and so on. The blending can be done rather crudely. Depositing materials in the raw coal bin by alternating truck loads will be sufficient. Proper control of coal logistics and analysis of coal blend have to be put in place.

5.3 Coal injection in the tuyeres

Coals are injected via lances into the tuyeres, and gasified and ignited in the raceway. The coal is in the raceway area only for a very short time (5 milliseconds) and so the characteristics of the gasification reaction are very important for the effectiveness of a PCI system. Coal gasification consists of several steps as outlined in Figure 5.3. Firstly the coal is heated and the moisture evaporates. Gasification of the volatile components then occurs after further heating. The volatile components are gasified and ignited, which causes an increase in the temperature. All of these steps occur sequentially with some overlap.

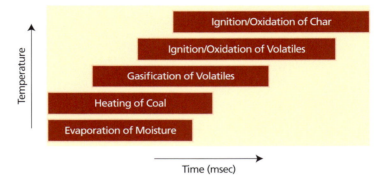

Figure 5.3   Coal gasification

The effects of lance design, extra oxygen and coal type on the coal combustion have been analysed. Originally, the coal lances were straight stainless steel lances that were positioned at or close to the tuyere/blowpipe interface as indicated in Figure 5.4 on the next page. Occasionally, very fine carbon formed from gas is detected as it leaves the furnace through the top. To avoid this problem, especially at high injection rates, companies have installed different types of injection systems at the tuyeres, such as:
– Coaxial lances with oxygen flow and coal flow.
– Specially designed lances with a special tip to get more turbulence at the lance tip.
– Use of two lances per tuyere.
– Bent lance tips, positioned more inwards in the tuyere.

When using PCI, deposits of coal ash are occasionally found at the lance tip or within the tuyere. The deposits can be removed by periodic purging of the lance by switching off the coal while maintaining air (or nitrogen) flow.

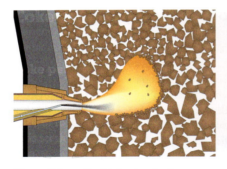 

Figure 5.4   Coal injection in the tuyeres and lance positioning

The speed of gasification increases as;
- The volatility of the coals increases.
- The size of the coal particles decreases.
- The blast and coal are mixed better. Moreover, as the injection level increases, the amount of coal that leaves the raceway without being gasified increases.

The gasification of coal also depends on the percentage of volatile matter (VM). If low volatile coals are used, a relatively high percentage of the coal is not gasified in the raceway and is transported with the gas to the active coke zone. This "char" will normally be used in the process, but might affect the gas distribution. The high volatile (HV, over 30 % VM) and ultra high volatile coal (over 40 % VM) produces a large quantity of gas in the raceway and a small quantity of char. If the gas combustion is not complete, soot can be formed. Blending a variety of injection coals, especially high volatile and low volatile coals, gives the advantage of being able to control these effects.
It has been found that the coke at the border between raceway and dead man contains more coke fines when working at (high) injection rates. This region has been termed the "bird's nest".

5.4   Process control with pulverised coal injection

5.4.1   Oxygen and PCI

At high Pulverised Coal Injection operation about 40% of the reductant is injected via the tuyeres. Therefore, it is important to control the amount of coal per tonne hot metal as accurate as the coke rate is controlled. The feed tanks of the coal injection are weighed continuously and the flow rate of the coal is controlled. It can be done with nitrogen pressure in the feed tanks or a screw or rotating valve dosing system. In order to calculate a proper flow rate of coal (in kg/minute) the hot metal production has to be known. There are several ways to calculate the production. The production rate can be derived from the amount of material charged into the furnace. Short–term corrections can be made by calculating the oxygen consumption per tonne hot metal from the blast parameters in a stable period and then calculating the actual production from blast data. Systematic errors and/or the requirement for extra coal can be put in the control model.

The heat requirement of the lower furnace is a special topic when using PCI. Coal is not only used for producing the reduction gases, but use of coal has an effect on the heat balance in the lower furnace. The heat of the bosh gas has to be sufficient to melt the burden: define the "melting heat" as the heat needed to melt the burden. The heat requirement of the burden is determined by the "pre–reduction degree", or how much oxygen has still to be removed from the burden when melting. The removal of this oxygen requires a lot of energy. The "melting capacity" of the gas is defined as the heat available with the bosh gas at a temperature over 1500 °C. The melting capacity of the gas depends on:
– The quantity of tuyere gas available per tonne hot metal. Especially when using high volatile coal there is a high amount of H_2 in the bosh gas.
– The flame temperature in the raceway.

The flame temperature in itself is determined by coal rate, coal type, blast temperature, blast moisture and oxygen enrichment.

From the above, the oxygen percentage in the blast can be used to balance the heat requirements of the upper and lower furnace. The balance is dependent on the local situation. It depends e.g. on burden and coke quality and coal type used. For the balance there are some technical and technological limitations, which are presented as an example in Figure 5.5. For higher injection rates more oxygen is required. The limitations are given by:
– Too low top gas temperature. If top gas temperature becomes too low it takes too long for the burden to dry and the effective height of the blast furnace shortens.
– Too high flame temperature. If flame temperature becomes too high burden descent can become erratic.
– Too low flame temperature. Low flame temperature will hamper coal gasification and melting of the ore burden.
– Technical limitations to the allowed or available oxygen enrichment.

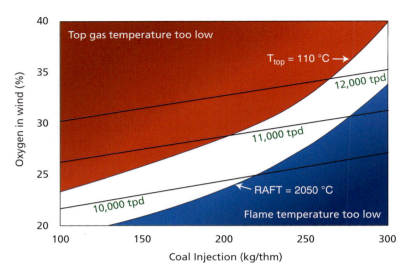

Figure 5.5   Limiting factors affecting raceway conditions with Pulverised Coal Injection (RAFT = Raceway Adiabatic Flame Temperature)

The higher the oxygen injection, the higher the productivity of the furnace as shown in Figure 5.5, which is an example based on mass and heat balance of an operating furnace. The highest productivity is reached, with an oxygen level, so that the top gas temperature is at the minimum. The minimum is the level, where all all water of coke, burden and process is eliminated from the furnace, i.e. slightly above 100 °C. From a technological perspective it can be said, that the heat balances over the lower part of the furnace (i.e. from 900 °C to tuyere level) and over the upper part of the furnace (i.e. from top to the 900 °C isotherm) are in balance (Section 8.5).

5.4.2   Effect of additional PCI

The effect of the use of extra coal injection for recovery of a cooling furnace is twofold. By putting extra coal on the furnace the production rate decreases. Simultaneously, the flame temperature drops. If the chilling furnace has insufficient melting capacity of the gas, extra PCI may worsen the situation. In such a situation the efficiency of the process must be improved, i.e. by a lower production rate and lower blast volume.

This is illustrated in Table 5.1. The table shows that additional coal injection slows down the production rate, because the coke burning rate decreases. It is a typical example; the precise effect depends on coke rate and coal type used. A furnace recovers from a cold condition by increasing PCI, *because* it slows down the production rate. If, however, the flame temperature is relatively low, the effect of the drop in flame temperature can be as large as the effect of the decreased production rate.

Starting Situation	
Operating parameters	
Coke rate	300 kg/tHM
Coal injection rate	200 kg/tHM
Replacement ratio	0.85 kg coal/kg coke
Flame temperature	2,200 °C
Coke and coal consumption in normal operation (as kg standard coke/tHM)	
Coke introduced	300
Coal introduced	170
Total coke and coal	470
Consumption to be subtracted to determine burn rates:	
Carbon in hot metal	−50
Direct reduction	−120
Result: total burn rate in front of tuyeres	**300**
of which coal	170
and thus coke	130
Changed situation if an additional 10 kg/tHM of coal is injected	
Total burn rate remains	**300**
of which coal	178.5
and thus coke	121.5
Production rate decrease (fully determined by coke burn rate)	6.5%
Flame temperature drop	32 °C
Gas melting capacity drop (heat > 1,500 °C)	4.6%

Table 5.1   Effect of additional coal injection

5.5 Circumferential symmetry of injection

If every tuyere in a blast furnace is considered as part of the blast furnace pie and is responsible for the process to the stock–line, it is self evident that the circumferential symmetry of the process has to be assured to reach good, high performance. The various systems in use for PCI have different methods to ensure a good distribution.

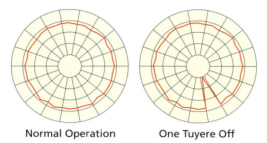

Normal Operation   One Tuyere Off

Figure 5.6   Schematic presentation of the effect of no PCI on one tuyere

PCI at all tuyeres		PCI at one tuyere off	
• Coke Rate	300 kg/t	• Burns at tuyeres	3300 kg/hr
• PCI 200 (RR =0,85)	170 kg/t	• All coke	
• Total	470 kg/t		
• Production	10 t/hr	Production increase at this tuyere without PCI of 3300/1300 = 254%!	
• Carbon balance:			
• Coke	3000 kg/hr		
• Coal (in SRE)	1700 kg/h		
• Total	4700 kg/hr		
• Iron carbonization	-500 kg/hr		
• To direct reduction	-1200 kg/hr		
• Burns at tuyeres	3000 kg/hr		
• Of which Coal	1700 kg/hr		
• and Coke	1300 kg/hr		

Table 5.2   Coke use per tuyere in case a single tuyere receives no coal

However, the largest deviation from circumferential symmetry occurs when no coal is injected in a particular tuyere. If no injection is applied, the production rate at that particular tuyere increases substantially. Consequently, the blast furnace operator has to take care that all tuyeres are injecting coal. In particular, where two tuyeres next to each other are not injecting coal the equalising effects between the tuyeres are challenged. Especially if the furnace is operating at high PCI rates, the situation is rather serious and short–term actions have to be taken to correct the situation.

This point can be illustrated from Table 5.2 and Figure 5.6. The calculation shows, how much coke is consumed in front of a tuyere, where coal injection is switched off. At high injection rates, the production can increase twofold or more. Note, that this is an example, since in such a situation neighbouring tuyeres will tend to contribute. Moreover, the calculation does not take the oxygen of the coal itself into account.

With coal injection it is very important that the tuyeres are clear and open, allowing the coal plume to flow into the raceway. If the tuyere should become blocked, or a blockage in front of the tuyere appears, then the coal must be removed immediately. If it is not, then the coal will be forced backwards into the tuyere stock and can ignite further up in the connection with the bustle pipe (see Figure 5.7). This can cause serious damage or even explosions. The phenomenon has also been observed with natural gas injection.

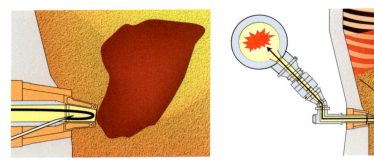

Figure 5.7 Coal backing up into the bustle pipe, caused by scab in front of tuyere, leading to possibility for explosion

To prevent this, a light sensor may be fitted in front of the peep-sight to detect a blockage at the end of the tuyere, or the delta-P can be measured over the tuyere to detect when flow has stopped, indicating that a blockage is present. The coal to that tuyere is automatically switched off and restarted only once an operator has checked to see if the tuyere can accept coal.

5.6 Gas and oil injectants

As stated earlier, all types of (hydrocarbon) injectants can be used. A comparison of replacement ratio, typical chemical composition and effect on flame temperature are given in Table 5.3.

Injectant	Replacement Ratio*	C %	H %	Moisture %	Effect flame temp. °C**
Coal	0.80	78–82	4–5.5	1–4	–32
Oil	1.17	87	11	2	–37
Natural gas	1.05	57	19	—	–45
Tar	1.0	87	6	2	–25

Table 5.3 Typical data for injectants
*) Compared with standard coke with 87.5% C
**) When injecting additional 10 kg/tHM

VI Burden Calculation and Mass Balances

6.1 Introduction

The blast furnace is charged with pellets, sinter, lump ore and coke, while additional reductant might be injected through the tuyeres. The steel plant requires a defined quality of hot metal and the slag has to be chosen for optimum properties with respect to fluidity, desulphurising capacity and so on. Therefore, the blast furnace operator has to make calculations to select the blast furnace burden. The present chapter first indicates the conditions for a burden calculation, which is then illustrated with a practical example. Later in the chapter the burden calculation is taken a step further to indicate the process results. To this end a simple one–stage mass balance is used.

6.2 Burden calculation: starting points

Starting points for burden calculations are the hot metal and slag quality.
– Hot metal quality: silicon, typically 0.4 to 0.5 %. Low sulphur (under 0.03 %) and defined phosphorous levels, which vary due to variation in burden materials from 0.05 to 0.13 %.
– Slag quality: generally the lower the slag volume the better. Typically the four major constituents of slag contain about 96% of the total volume: Al_2O_3 (8 to 20 %), MgO (6 to 12 %), SiO_2 (28 to 38 %) and CaO (34 to 42 %). For slag design, see Chapter X.

The burden has to fulfil requirements with respect to:
– Maximum phosphorous input, since phosphorous leaves the furnace with the iron.
– Maximum alkali input, especially potassium, which can attack the refractory and affect the process. Typically a limit of 1 to 1.5 kg/tHM is used.
– Maximum zinc input: zinc can condense in the furnace and can, similar to alkali, lead to a Zn cycle.

Typically, limits for zinc input are 100–150 g/tHM. With high central gas temperatures, zinc and alkali are partly removed with the top gas.

6.3 An example of a burden calculation

The burden calculation uses the chemical composition (on a dry basis) and the weights of the various materials in a charge as input parameters. A charge consists of a layer of burden material and coke with its auxiliary reductants as injected through the tuyeres. In order to be able to do the calculation, the yield losses when charging the furnace are also taken into account. The present example is restricted to the components required to calculate the slag composition. The four main components (SiO_2, CaO, MgO and Al_2O_3) represent 96 % of the total slag volume. The other 4 % consist of MnO, S, K_2O, P and many more. The losses from the materials charged through the top into the blast furnace are taken into account and are normally based on samples of material from the dust catcher and scrubber systems. The calculation is presented in Table 6.1.

	Chemical analysis							
	Ash	Moisture	Loss	Fe	SiO_2	CaO	MgO	Al_2O_3
Coke	9 %	5 %	2 %	0.5 %	5.0 %			3.0 %
Coal	6 %	1 %	0 %	0.2 %	3.0 %			1.5 %
Sinter		1 %	1 %	58 %	4.0 %	8.3 %	1.4 %	0.6 %
Pellets		1 %	1 %	65 %	3.5 %		1.3 %	0.8 %
Lump		3 %	1 %	61 %	4.0 %			1.0 %
Burden								
	Weight kg/tHM	After losses kg/tHM		Input kg/tHM				
Coke	300	294		1	15	0	0	9
Coal	200	200		0	6	0	0	3
Sinter	1000	990		575	40	82	14	6
Pellets	500	495		322	17	0	6	4
Lump	80	79		49	3	0	0	1
Total	1580			947	81	82	20	23
Correction: HM silicon 0.46 % = 10 kg SiO_2/tHM					−10			
Slag					71	82	20	23
Results								
Slag volume *)			kg/tHM	204	SiO_2	CaO	MgO	Al_2O_3
Slag composition					35 %	40 %	10 %	11 %
Basicity	CaO/SiO_2		1.16					
	$(CaO+MgO)/SiO_2$		1.45					
	$(CaO+MgO)/(SiO_2+Al_2O_3)$		1.10					
	Al_2O_3		11%					
	Ore/coke ratio		5.3					

*) (SiO_2+CaO+MgO+Al_2O_3)/0.96

Table 6.1   Simplified Burden Calculation

6.4 Process calculations: a simplified mass balance

The calculations of the previous section can be extended to include the blast into the furnace. In doing so the output of the furnace can be calculated: not only the hot metal and slag composition and the reductant rate, but the composition of the top gas as well. Calculation of the top gas composition is done in a stepwise manner in which the balances of the gas components (nitrogen, hydrogen, oxygen, CO and CO_2) and iron and carbon are made. For the calculations the example of a 10,000 t/d furnace is used. The stepwise approach indicated in Table 6.2.

Input Element	Nitrogen (N_2)	Hydrogen (H_2)	Iron (Fe)	Carbon (C)	Oxygen (O_2)
Main Sources	Blast	Injection Blast Moisture	Burden	Coke Injection	Burden (52 %) Blast (48 %)
What to know	N_2 % in blast	H % in reductant	%Fe ore burden	%C in coke and injectant	% O_2 wind
Main output via	Top gas	Top gas	Hot metal	Top gas (85%) Hot metal (15%)	Top Gas – CO (32 %) – CO_2 (64 %) – H_2O (4 %)
What to know	N_2 % in top gas	H_2 efficiency	Hot metal composition	Rates per tonne Composition	
Calculation of	Top gas volume	H_2 % in top gas	Oxygen input via burden	Top gas composition CO & CO_2 %	

Table 6.2   Stepwise approach for a simplified mass balance

The approach is as follows:

Step 1: nitrogen balance: from the nitrogen balance the total top gas volume is estimated.

Step 2: hydrogen balance: from hydrogen input and a hydrogen utilisation of 40 % the top gas hydrogen can be estimated. In practice hydrogen utilisations of 38–42 % are found.

Step 3: iron and carbon balance: the carbon use per tonne is known from the hot metal chemical composition and coke and coal use per tonne.

Step 4: oxygen balance: the burden composition gives the amount of oxygen per tonne hot metal input at the top, while also the amount of oxygen with the blast is also known per tonne hot metal.

Step 5: the balances can be combined to calculate the top gas composition.

The calculations are based on basic chemical calculations. Starting points for the calculations are, that:
- 12 kilogram of carbon is a defined number of carbon atoms defined as a kilomole.
- Every mole of an element or compound has a certain weight defined by the periodic table of the elements.

– 1 kmole of a gas at atmospheric pressure and 0 °C occupies 22.4 m³ STP. The properties of the various components used for the calculations are indicated in Table 6.3. The present balance is used for educational purposes figures and compositions are rounded numbers. Effects of moisture in pulverised coal and the argon in the blast are neglected.

	Atomic weight			Molecular weight	
N_2	28	kg/kmole	CO	28	kg/kmole
O_2	32	kg/kmole	CO_2	44	kg/kmole
H_2	2	kg/kmole			
C	12	kg/kmole			
Fe	55.6	kg/kmole			
Si	28	kg/kmole			

Table 6.3 Properties of materials used for mass balance calculations
1 kmol gas (N_2, O_2, etc) = 22.4 m³ STP
1 tonne hot metal contains 945 kg Fe= 945/55.6 = 17.0 kmole

6.4.1 The nitrogen balance

Nitrogen does not react in the blast furnace, so it escapes unchanged via the top gas. At steady state the input equals the output and the top gas volume can be calculated with a nitrogen balance given the nitrogen input and the nitrogen concentration in the top gas. The input data for a simplified model are shown in Table 6.4 and the top gas volume is calculated in Table 6.5.

Blast volume	6500	m³ STP/min		
Oxygen in blast	25.6	%		
Moisture	10	g/m³ STP		
Production	6.9	tHM/min		
Coal rate	200	kg/tHM		
Coke rate	300	kg/tHM		
Top gas				
CO2	22	vol%		
CO	25	vol%		
H2	4.5	vol%		
N2	48.5	vol%		
	C%	H%	O%	N%
Coal	78	4.5	7	1.4
Coke	87	0.2		1.4

Table 6.4 Mass Balance Input

Burden Calculation and Mass Balances

Nitrogen from blast	(1−0.256)× 6500	4836	m³ STP/min
From coal		16	m³ STP/min
From coke		23	m³ STP/min
Total input		4875	m³ STP/min
Top gas nitrogen		48.5	%
Top gas volume		10051	m³ STP/min

Table 6.5   The nitrogen balance and top gas volume

6.4.2 The hydrogen balance

Moisture in the blast and coal reacts to H_2 and CO according to:
$$H_2O + C \rightarrow H_2 + CO$$

All hydrogen in coal and coke are converted to H_2 in the furnace. In the furnace the H_2 is reacting to H_2O; part of the hydrogen is utilised again. Since the top gas volume is known as well as the hydrogen input, the top gas hydrogen can be calculated, if a utilisation of 40% is assumed. There are ways to check the hydrogen utilisation, but it is beyond the scope of the present exercise. Table 6.6 shows the input and calculates the top gas hydrogen.

	kg/min	in m³ STP/min
From blast	7	
From coal	56	
From coke	4	
Total input	67	750
Utilisation 40%, so 60% left in top gas		450
Top gas volume		10051
H_2 in top gas		4.5%

Table 6.6   The Hydrogen Balance

6.4.3 The iron and carbon balance

Hot metal contains 945 kg Fe per tonne. The balance is taken by carbon (45 kg), silicon, manganese, sulphur, phosphorous, titanium and so on. The precise Fe content of hot metal depends slightly on the thermal state of the furnace and quality of the input. For the balance we use 945 kg Fe/tHM. This amounts to 17 kmole (947/55.6).

The carbon balance is more complicated. The carbon is consumed in front of the tuyeres and is used during the direct reduction reaction (see section 8.2.1). The carbon leaves the furnace via the top gas and with the iron. The carbon balance is made per tonne hot metal. Table 6.7 shows the results. The carbon via the top gas is also given in katom per tonne hot metal.

Carbon used	In kg/tHM	katom/tHM
Carbon from coke	261	
Carbon from coal	156	
Total carbon use	417	
Carbon via iron	−45	
Carbon via top gas	372	31.0

Table 6.7   The Carbon Balance

6.4.4 The oxygen balance

The oxygen balance is even more complicated. Oxygen is brought into the furnace with the blast, with PCI, with moisture and with the burden. It leaves the furnace through the top. The burden with sinter contains not only Fe_2O_3 (O/Fe ratio 1.5) but some Fe_3O_4 (O/Fe ratio 1.33) as well. The O/Fe ratio used here is 1.46, which means that for every atom of Fe there is 1.46 atom O. On a weight basis it means, that for every tonne hot metal, which contains 945 kg Fe atoms there is 397 kg O–atoms. All this oxygen leaves the furnace with the topgas. The balance is given in Table 6.8.

		m³ STP /tHM	kg O/tHM	Katom O/tHM
Input	From blast	240	342	
	From blast moisture		8	
	From coal		14	
	From burden		397	
Total input			762	
Output via top gas			762	47.6

Table 6.8   The Oxygen Balance

6.4.5 Calculation of top gas analysis

The oxygen in the top gas is leaving the furnace in three different states:
– Bound to the hydrogen. The quantity is known since we know how much hydrogen has been converted to process water.
– Bound to carbon as CO.
– Bound to carbon as CO_2.

From the combination of the carbon balance and the oxygen balance we can now derive the top gas utilisation, as shown in Table 6.9.

	Katom/tHM
Carbon via top gas	31.0
Oxygen via top gas	47.6
Oxygen bound to hydrogen	−1.9
Oxygen as CO and CO_2	45.7

Oxygen balance: CO + 2x CO_2		45.7
Carbon balance: CO + CO_2		31.0
CO_2		14.7
CO		16.3
Utilisation	$CO_2/(CO+CO_2)$	47.3 %
CO_2 volume	2283	m³ STP/min CO_2 % 22.7 %
CO	2539	m³ STP/min CO % 25.3 %

Table 6.9   Calculation of Top Gas Utilisation

The calculations can be used to check the correct input data. More advanced models are available, which take into account the heat balance of the chemical reactions as well (e.g. Rist and Meysson, 1966). The models are useful for analysis, especially questions like "are we producing efficiently?" and for prediction: what if PCI is increased? hot blast temperature is increased? and so on.

VII The Process: Burden Descent and Gas Flow Control

7.1 Burden descent: where is voidage created?

The burden descends in the blast furnace from top to bottom. Figure 7.1 shows a representation of the burden descent. It is indicated with stock rods, which are resting on the burden surface and descending with the burden between charging. The burden surface descends with a speed of 8 to 15 cm/minute.

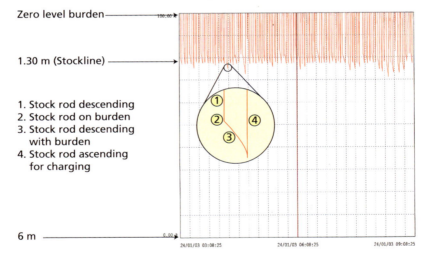

Figure 7.1   Stable burden descent

In order for the burden to descend, voidage has to be created somewhere in the furnace. Where is this voidage created? See Figure 7.2.
– Firstly, coke is gasified in front of the tuyeres, thus creating voidage at the tuyeres.
– Secondly, the hot gas ascends up the furnace and melts the burden material. So the burden volume is disappearing into the melting zone.
– Thirdly, the dripping hot metal consumes carbon. It is used for carburisation of the iron as well as for the direct reduction reactions, so below the melting zone coke is consumed.

It is possible to indicate how much each of the three mechanisms contributes to the amount of voidage created. A large part of the voidage is created at the melting zone. In a typical blast furnace on high PCI, only about 25 % of the voidage is created at the tuyeres.

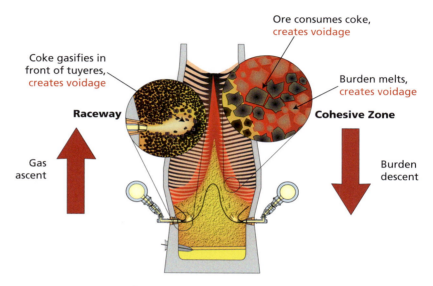

Figure 7.2   Creation of voidage in the Blast Furnace

This implies that the mass flow of material is strengthened towards the ring where the highest amount of ore is charged into the furnace. Therefore, at low coke rates high ore concentration at any ring in the circumference, especially in the wall area, has to be avoided.

Sometimes the burden descent of a blast furnace is erratic. What is the mechanism? Ore burden materials and coke flow rather easily through bins, as can be observed in the stock house of a blast furnace. Hence in the area in the blast furnace where the material is solid, the ore burden and coke flow with similar ease to the void areas. Nevertheless, blast furnace operators are familiar with poorly descending burden (Figure 7.3). Also the phenomenon of "hanging" (no burden descent) and "slips" (fast uncontrolled burden descent) are familiar. From the analysis in this section it follows that, in general, the cause of poor burden descent must be found in the configuration of the melting zone. The materials "glue" together and can form internal bridges within the furnace. Poor burden descent arises at the cohesive zone. The effect of a slip is, that the layer structure within the furnace is disrupted and the permeability for gas flow deteriorates (See Figure 7.22).

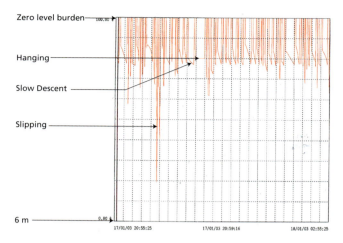

Figure 7.3   Irregular Burden Descent

7.2 Burden descent: system of vertical forces

The burden descends because the downward forces of the burden exceed counteracting upward forces. The most important downward force is the weight of the burden; the most important upward force is the pressure difference between the blast and top pressure.

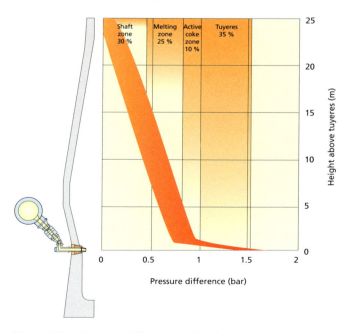

Figure 7.4   Pressure difference over burden

The cohesive zone is the area with the highest resistance to gas flow, which leads to a high pressure drop over the cohesive zone and to a large upward force. If this pressure difference becomes too high, the burden descent can be disturbed. This happens for instance, when a blast furnace is driven to its limits and exceeds the maximal allowable pressure difference over the burden.

In addition to the upward force arising from the blast pressure, friction forces from the descending burden are impacting on the burden descent: the coke and burden are pushed outward over a cone of stationary or slowly descending central coke. Also the wall area exerts friction forces on the burden. In case of irregular burden descent these friction forces can become rather large.

The coke submerged in hot metal also exerts a high upward force on the burden due to buoyancy forces (Figure 7.5) as long as the coke is free to move upwards and does not adhere to the bottom.

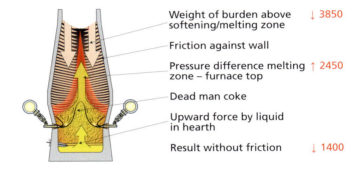

Figure 7.5   System of vertical forces in the Blast Furnace

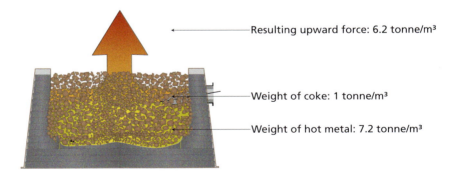

Figure 7.6   Upward force from hearth liquids

In operational practice poor burden descent is often an indicator of a poor blast furnace process. The reasons can be:
– The upward force is too high. Experienced operators are well aware of the maximum pressure difference over the burden that allows smooth operation. If the maximum allowable pressure difference is exceeded (generally 1.6 to 1.9

bar), the process is pushed beyond its capabilities: burden descent will become erratic, resulting in frequent hanging, slipping and chills.
– A hot furnace is also known to have poorer burden descent. This is because the downward force decreases due to the smaller weight of burden above the melting zone. In addition, there is more slag hold–up above the tuyeres, because of the longer distance and the (primary) slag properties.
– Burden descent can be very sensitive to casthouse operation because of the above–mentioned upward force on the submerged coke.

7.3 Gas flow in the blast furnace

The gas generated at the tuyeres and at the melting zone has a short residence time of 6 to 12 seconds in the blast furnace (section 2.3). During this time the gas cools down from the flame temperature to the top gas temperature, from 2000 to 2200 °C down to 100 to 150 °C, while simultaneously removing oxygen from the burden. The vertical distance between tuyeres and stockline is around 22 metres. Therefore, the gas velocity in the furnace is rather limited, in a vertical direction about 2 to 5 m/s, which is comparable with a wind speed of 2 to 3 Beaufort, during the 6 to 12 seconds the chemical reactions take place.

How is the gas distributed through the furnace? First consider the difference between the coke layers and the ore burden. It is important to note, as indicated in Figure 7.7, that ore burden has a higher resistance to gas flow than coke. The resistance profile of the furnace determines how gas flows through the furnace. The gas flow along the wall can be derived from heat losses or hot face temperatures as the gas will heat the wall as it travels past.

	Voidage	Diameter
Ore Burden	low	small
Coke	high	large

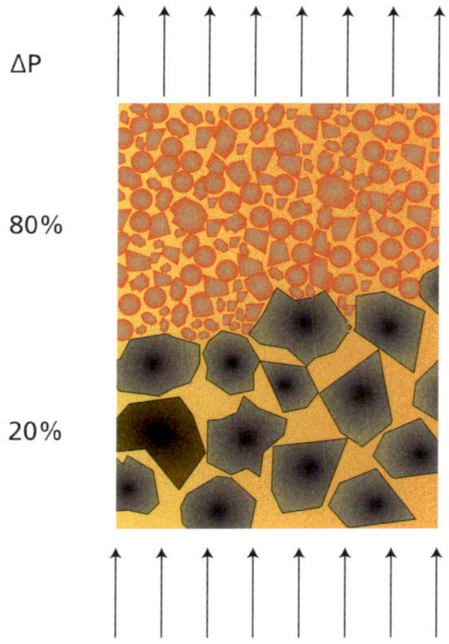

Figure 7.7   Pressure loss through coke and ore

As soon as the ore burden starts to soften and melt at about 1100 °C, the burden layer collapses and becomes (nearly) impermeable for gas. If this happens in the centre of the furnace the central gas flow is blocked.

7.3.1 Observation of heat fluxes through the wall

Figure 7.8 shows the temperature at the hot face of the furnace wall. It has been observed in many furnaces, that suddenly the temperature rises in minutes and decreases over the next hour(s). This is often attributed to the loss of scabs (build– up) on the furnace wall. The explanation put forward in this book is that such temperature excursions are due to "short–circuiting" of gas along the furnace wall. These "short–circuits" are due to the formation of gaps along the furnace wall creating a very permeable area where the hot gasses preferentially flow. This can be observed from pressure tap measurements (see Figure 7.25). Low CO_2 concentrations in the wall area during such excursions have been observed and confirm the "short–circuiting". The basic premise of the present book is that heat losses through the wall are caused by gas flow along the walls. The gas is more or less directly coming from the raceway.

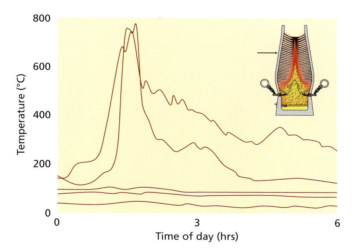

Figure 7.8   *Temperatures at hot face*

Why does the gas flow along the wall? Gas takes the route with the lowest resistance and therefore highest permeability. The resistance for gas flow in a filled blast furnace is located in the ore layers, since its initial permeability is 4 to 5 times less than the permeability of coke layers. There are two areas in the blast furnace that have the highest permeability: the centre of the furnace if it contains sufficient coke and the wall area. At the wall there can be gaps between the descending burden and the wall. In the centre of the furnace there can be a high percentage of coke and there can be relatively coarse ore burden due to segregation.

7.3.2 Two basic types of cohesive zone

The efficiency of the furnace is determined by the amount of energy used in the process. Heat losses to the wall and excess top gas temperature are examples of energy losses. The top gas contains CO and H_2, which have a high calorific value, therefore, the efficiency of a blast furnace is determined by the progress of the chemical reactions and thus by the gas flow through the furnace.

Two basic types of gas distribution can be discriminated: the "central working" furnace and the "wall working" furnace. The typology has been developed to explain differences in operation. Intermediate patterns can also be observed. In the "central working" furnace the gas flow is directed towards the centre. In this case the centre of the furnace contains only coke and coarse burden materials and is the most permeable area in the furnace. The cohesive zone takes on an "inverted V shape". In a "wall working" furnace the gas flow through the centre is impeded, e.g. by softening and melting burden material. The gas flows preferentially through the zone with highest permeability, i.e. the wall zone. In this case the cohesive zone takes the form of "W shape". Figure 7.9 shows both types.

Both types of gas flow can be used to operate a blast furnace, but have their own drawbacks. The gas flow control is achieved with burden distribution.

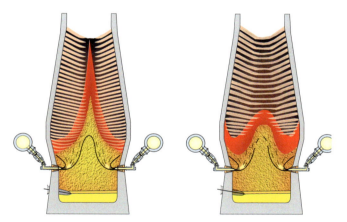

Figure 7.9   Two types of melting zone, Central working (left) and Wall–working (right)

7.3.3 Central working furnace

The two types of gas flow through a furnace can be achieved with the help of the burden distribution. In Figure 7.10 the ore to coke ratio over the radius is shown for a central working furnace. In the figure the centre of the furnace only contains coke. Therefore, in the centre of the furnace no melting zone can be formed and the gas is distributed via the coke slits from the centre towards outside radius of the furnace. The melting zone gets an inverted V or even U shape. The central coke column not only serves as a gas distributor, but as well as a type of pressure valve: it functions to stabilise the blast pressure.

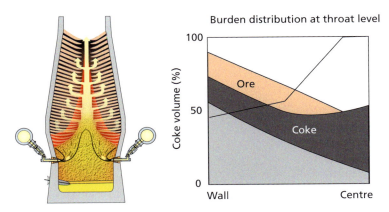

Figure 7.10   Central working furnace

It depends on the type of burden distribution equipment how the coke can be brought to the centre. With a bell–less top the most inward positions of the chute can be used. With a double bell system the coke has to be brought to the centre by coke push (see below) and by choosing the right ore layer thickness in order to prevent the flooding of the centre with ore burden materials. In the central working furnace there is a relatively small amount of hot gas at the furnace wall: hence low heat losses. As a result the melting of the burden in the wall area takes place close to the tuyeres, so the root of the melting zone is low in the furnace. The risk of this type of process is that ore burden is not melted completely before it passes the tuyeres. This could lead to the observation of lumps of softened ore burden through the tuyere peep sites. This can lead from slight chilling of the furnace (by increased direct reduction) and irregular hot metal quality to severe chills and damage of the tuyeres.

Limiting the risk of a low melting zone root can be done with gas and burden distribution. Operational measures include the following.
– Maintain a sufficiently high coke percentage at the wall. Using nut coke in the wall area can also do this. Note that an ore layer of 55 cm at the throat needs about 20 to 22 cm of coke for the carburisation and direct reduction. So if the coke percentage at the wall is under 27 %, a continuous ore burden column can be made at the wall.
– Ensure a minimum gas flow along the wall in bosh and belly, which can be monitored from heat loss measurements and/or temperature readings. If the gas flow along the wall becomes too small, it can be increased by means of burden distribution (more coke to the wall or less central gas flow) or by increasing the gas volume per tonne hot metal (by decreasing oxygen).

– Control the central gas flow. Note that the gas flow through the centre leaves the furnace at a high percentage of CO and H_2 and a high temperature. The energy content of the central gas is not efficiently used in the process and thus the central gas flow should be kept within limits.

The central working furnace can give very good, stable process results with respect to productivity, hot metal quality and reductant rate. It also leads to long campaign length for the furnace above the tuyeres. However, the process is very sensitive for deviations in burden materials, especially the size distribution.

7.3.4 Wall working furnace

In Figure 7.11 the wall working furnace is presented. Melting ore burden blocks the centre of the furnace and the gas flow is directed towards the wall area.

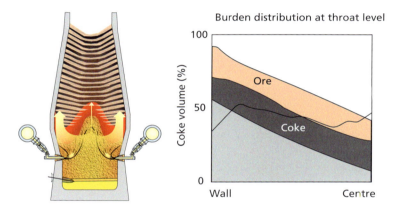

Figure 7.11   Wall working furnace

The gas flow causes high heat losses in the area of the furnace where a gap can be formed between burden and wall i.e. in lower and middle shaft. The melting zone gets a W shape or even the shape of a disk. In this situation the root of the melting zone is higher above the tuyeres, which makes the process less sensitive for inconsistencies. The process can be rather efficient. However, due to the high heat losses the wear of the refractory in the shaft is much more pronounced than with the central working furnace. The gas passing along the wall can also cool down rapidly and in doing so loses its reduction capabilities. As a consequence, the fuel rate is high. Moreover the fluctuations in the pressure difference over the burden are more pronounced, which leads to limitations in productivity.

7.3.5 Gas distribution to ore layers

Gas produced in the raceway is distributed through the coke layers in the cohesive zone and into the granular coke and ore layers, as shown in Figure 7.12.

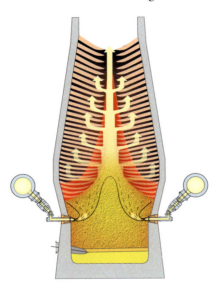

Figure 7.12 Schematic presentation of gas distribution through coke layers

The ore burden layers account initially for about 80% of the resistance to gas flow. The reduction process takes place within these layers.

What determines the contact between the gas and the ore burden layers? The most important factor determining the permeability to gas flow is the voidage between particles. As mentioned in Section 3.2.1 the voidage between particles depends heavily on the ratio of coarse to small particles. The wider the size distribution, the lower the voidage. Moreover, the finer the materials, the lower the permeability (Chapter 3). In practical operations the permeability of ore burden material is determined by the amount of fines (percentage under 5 mm). Fines are very unevenly distributed over the radius of the furnace, as is indicated by the typical example shown in Figure 7.13. Fines are concentrated along the wall especially under the point of impact of the new charge with the stockline.

If a bell–less top is used, the points of impact can be distributed over the radius. With a double bell charging system the fines are concentrated in a narrow ring at the burden surface and close to the wall. When the burden is descending the coarser materials in the burden follow the wall, while the fines fill the holes between the larger particles and do not follow the wall to the same extent as the coarser particles. Therefore, upon descent the fines in the burden tend to concentrate even more.

Moreover, sinter and lump ore can break down during the first reduction step (from haematite to magnetite). This effect is stronger if the material is heated more slowly. Thus, the slower the material is heated the more fines are generated, the extra fines impede the gas flow even more, giving rise to even slower heating.

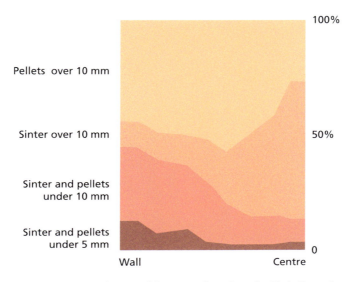

Figure 7.13   Distribution of fines over the radius, double bell simulation (after Geerdes et al, 1991)

In summary:
- The permeability of the ore burden is determined by the amount of fines.
- The amount of fines is determined by:
- The screening efficiency in the stock house.
- The physical degradation during transport and charging.
- The method of burden distribution used.
- The low temperature degradation properties of the burden. These effects cause a ring of burden material with poor permeability in many operating blast furnaces. This ring of material in particular is often difficult to reduce and to melt down. Sometimes, unmolten ore burden materials are visible as scabs through the peepsites of the tuyeres. The unmolten material can cause operational upsets like chilling the furnace or tuyere failures. It is a misunderstanding to think that these scabs consist of accretions fallen from the wall.

7.4 Fluidisation and channelling

The average gas speed above the burden is rather low, as shown in chapter 2. However, in a central working furnace the gas speed might locally reach 10 m/s or more especially in the centre of the furnace. This is well above theoretical gas velocities at which fluidisation can be observed (Figure 7.14). Coke fluidises much more easily than ore burden because of its lower density. It is believed that the ore burden secures the coke particles in the centre, nevertheless, if local gas speeds become too high, fluidisation may occur. Fluidisation of coke has been observed in operating furnaces as well as models of the furnace. It leads to a relatively open structure of coke. It has even been observed, that pellets on the border of fluidising coke "dive" into the coke layers.

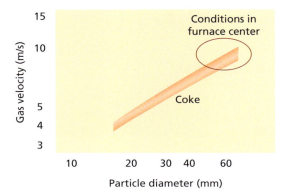

Figure 7.14   Gas velocities for fluidisation of ore burden and coke. Shaded areas indicate critical empty tube gas velocities for fluidization at 800 °C and 300 °C and 1 atmosphere pressure (after Biswas, 1981)

If the fluidisation stretches itself into the lower furnace, channelling can take place, short–circuiting the lower furnace (or even the raceway) with the top. These are open channels without coke or ore burden in it. Channelling is observed as a consequence of operational problems, for example, delayed casts can create higher local gas speeds, resulting in channelling. During channelling, the gas might escape through the top with a high temperature and low utilisation, since the gas was not in good contact with the burden. The limit of channeling is where the furnace slips.

7.5 Burden distribution

Burden distribution can be used to control the blast furnace gas flow. The conceptual framework of the use of burden distribution is rather complex, since the burden distribution is the consequence of the interaction of properties of the burden materials with the charging equipment.

7.5.1 Properties of burden materials

Figure 7.15 shows the angles of repose of the various materials used in a blast furnace. Coke has the steepest angle of repose, pellets have the lowest angle of repose and sinter is in between. Hence, in a pellet charged furnace the pellets have the tendency to slide to the centre.

Coke: 35–38 °   Sinter: 29–33 °   Pellets: 25–26 °

Figure 7.15   Segregation and angles of repose

Fines concentrate at the point of impact and the coarse particles flow "downhill" while the fine particles remain below the point of impact. This mechanism, known as segregation, is also illustrated in Figure 7.15.

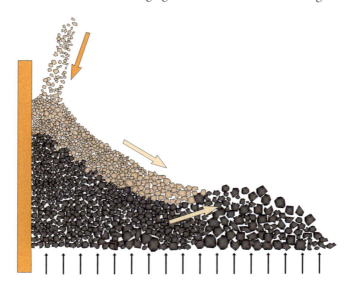

Figure 7.16   Coke push effect with gas flow

When burden is charged into the furnace, it pushes the coarse coke particles on the top of the coke layer towards the centre. This effect is called coke push and is more pronounced when the furnace is on blast. It is illustrated in Figure 7.16.

7.5.2 The charging equipment

The type of charging mechanism used has a major impact on the distribution of fines. Figure 7.17 shows the bell–less top and double bell systems.

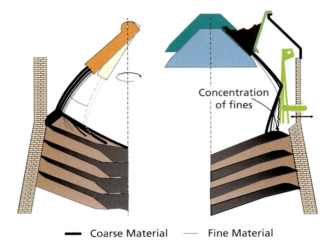

Figure 7.17   Bell–less top charging (left) and double bell charging (right): comparison of the segregation of fines on the stockline

In a bell–less top the possibility exists to distribute the fines in the burden over various points of impact by moving the chute to different vertical positions. Coke can be brought to the centre by programming of the charging cycle. With a double bell charging system there is less possibility to vary the points of impact and fines will be concentrated in narrower rings. Modern blast furnaces with a double bell charging system are mostly equipped with movable armour, which give certain flexibility with respect to distribution of fines and the ore to coke ratio over the diameter, especially at the wall. However, its flexibility is inferior to the more versatile bell–less system.

7.5.3 Mixed layer formation

The model of thinking applied up to here takes clean ore and coke layers as a starting point. However, since the average diameter of coke 45 to 55 mm is much larger than that of pellets and sinter (typically under 15 mm and 25 mm respectively) burden components dumped on a coke layer will tend to form a mixed layer (Figure 7.18). This mixed layer will have permeability comparable with the ore layer. The formation of mixed layers is also produced by protruding or recessed parts of the wall: such as protruding cooling plates, missing armour plates, wear of refractory at the throat and so on. The mixed layers have a different permeability and can give rise to circumferential process asymmetry. The smoother the burden descent, the less mixed layer formation.

7.5.4 Gas flow control

The optimised gas flow in a modern furnace operated at high productivity and low coke rate has the inverted V shaped melting zone type as described above. However, the gas escaping through the (ore–free) centre leaves the furnace with a low utilisation. This loss of "unused" gas should be minimised. If the central gas flow is too high, there is a too small gas flow along the wall for heating, reduction and melting of the ore burden and consequently the root of the melting zone comes close to the tuyeres. In this situation the reductant rate will be high and there is a high chance of tuyere damage. It is essential that the gas flowing though the centre distributes itself through the coke slits to the burden layers. Therefore, the permeability of the central coke column must not be too high, which means that the diameter of the central coke column must not be too wide. If the central gas flow is (partially) blocked, a relatively large part of the gas escapes along the wall and is cooled down low in the furnace. The reduction reactions slow down. In this situation the central gas flow is small and heat losses are high. Experience has shown that wall gas flow and central gas flow are strongly correlated. Gas flow control is based on keeping the balance between central and wall gas flow to the optimum.

The difficulty with gas flow control is that the gas flow is influenced by many changes in burden components, process parameters and installation specifics. The variation in the percentage of fines near (but not at) the wall and the low temperature breakdown properties of the burden are especially important.

The gas flow is closely monitored in order to control it. Instrumentation of the blast furnace is described in the next section.
The most important parameters to define the actual gas flow are:
– Burden descent (stock rods, pressure taps) and pressure difference over the burden.
– The wall heat losses or temperatures at the wall.
– Stockline gas composition and temperature profile.

Gas flow control and optimised burden distribution are found on a trial and error basis, and have to be developed for every furnace individually. Some general remarks can be made:
1. Gas flow is mainly controlled with coke to ore ratio over the radius. An example of a calculated burden distribution is shown in Figure 7.18. Note the ore free centre.
2. The centre of the furnace should be permeable and no or minimal (coarse) ore burden should be present.
3. The coke percentage at the wall should not be too low. Note that 70 cm of ore in the throat consumes about 25 cm of coke for direct reduction (Figure 7.19). A continuous vertical column of burden material should be prevented. A coke slit should be maintained between all ore layers.
4. Concentration of fines near the wall should be prevented.

5. The central gas flow is governed by the amount of ore burden reaching the centre. The amount of ore reaching the centre heavily depends on the ore layer thickness and the amount of coarse coke lumps. To reach a stable gas flow the central gas flow should be kept as consistent as possible and consequently, when changes in ore to coke ratio are required, the ore layer should be kept constant. This is especially important when changing the coal injection level as this will result in big changes in the relative layer thickness of ore and coke are made.
6. The coke layer thickness at the throat is typically in the range of 45 to 55 cm. In our example in section 2.3 it is 46 cm. The diameter of the belly is 1.4 to 1.5 times bigger than the diameter of the throat. Hence, the surface more than doubles during burden descent and the layer thickness is reduced to less than half the layer thickness at the throat. Japanese rules of thumb indicate that the layer thickness at the belly should not be less than 18 cm. The authors have, however, successfully worked with a layer thickness of coke at the belly of 14 cm.

In the practical situation small changes in ore layer thickness can strongly influence central gas flow. This effect is generally stronger in double bell–movable armour furnaces than in furnaces equipped with a bell–less top. An example for a burden distribution control scheme is given in Table 7.1. If more central gas flow is required then Coke 3 replaces schedule Coke 2. Replacing Coke 2 with Coke 1 reduces central gas flow.

	Position	11	10	9	8	7	6	5	4	3	2	1
		Wall							Centre			
Coke 1	More central	–	14 %	14 %	16 %	14 %	14 %	14 %	–	6 %	–	8 %
Coke 2	Normal	–	14 %	14 %	14 %	14 %	14 %	14 %	–	6 %	–	10 %
Coke 3	Less central	–	14 %	14 %	12 %	14 %	14 %	14 %	–	6 %	–	12 %
Ore		16 %	16 %	16 %	12 %	10 %	10 %	10 %	10 %			

Table 7.1 Bell–less top charging schedules with varying central gas flow

Similar schedules can be developed for a double bell charging system. With a double bell system, the use of ore layer thickness can also be applied: a smaller ore layer gives higher central gas flow and vice versa. If a major change in coke rate is required, the operator has the choice either to change the ore base and keep the coke base constant, or change the coke base and keep the ore base constant. Both philosophies have been successfully applied. The operators keeping the coke base constant point to the essential role of coke for maintaining blast furnace permeability, especially the coke slits. The authors, however, favour a system in which the ore base is kept constant. The gas distribution is governed by the resistance pattern of the ore burden layers and—as mentioned above—by the amount of ore burden that reaches the centre. The latter can change substantially when changing the ore base, especially in furnaces equipped with double bell charging. An illustrative example showing a change in coke rate from 350 kg/tHM to 300 kg/ tHM is presented in

Table 7.2. The ore base is kept constant and coke base reduced. Experience has shown that relatively minor changes in burden distribution will be required for optimisation of the central gas flow (i.e. coke distribution). The burden distribution adjustments can be applied as a second step if required.

	Old situation	New Situation
Coke rate	350 kg/tHM	300 kg/tHM
Coke base	21 t	18 t
Ore base	90 t	90 t
Burden distribution		No change until required

Table 7.2 Coke base change when PCI rate changes

Burden distribution changes should be based on an analysis of the causes of changes in gas flow. The gas flow can also be influenced by operational problems, such as a low burden level or problems in the casthouse. In this situation adjustments in the burden distribution will not give satisfactory results. Heat losses through the wall are very closely related to burden descent. Therefore, the cause of high heat loads should be analysed together with other process data. An example of a burden distribution is shown in Figure 7.18.

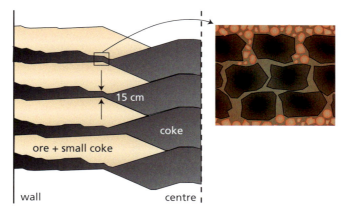

Figure 7.18 Example of burden distribution with an ore–free centre and ore burden penetration in coke layer

7.6 Coke layer

7.6.1 Coke percentage at wall

For optimum gas distribution through the coke layers it is desirable to have an ore–free chimney in the centre of the furnace. This then requires a large amount of coke to be present in the centre, but still some coke is required at the wall. This section deals with the question as to how much coke is required at the wall area.

A 70 cm thick ore layer at the wall contains about 1.5 tonnes ore burden in one square metre and therefore about 1 tonne hot metal. As shown in Section 8.2.1 dealing with direct reduction, the ore burden consumes coke, at a rate of about 120 kg coke per tonne. This amount of coke corresponds to a layer thickness of 24 cm, so the minimum coke amount at the wall is about 25% of the volume, (see Figure 7.19), assuming that the coke is used only for direct reduction.

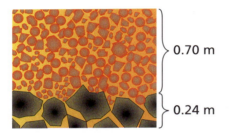

- 1 tHM is produced with 1.55 t ore
- 1.55 t ore is contained in a 1x1x0.70 m³ volume
- The 120 kg coke, required for direct reduction is contained in 1x1x0.24 m³

Corresponds to 0.70 m and 0.24 m thick layers

Figure 7.19   Coke required for direct reduction

If the amount of coke at the wall is less than the 25% of the volume, then the ore layers will make contact between the sequential layers upon melting. This will form a column of unmolten ore that descends down the furnace to the tuyeres. This will lead to disturbed gas flow, but also there is a risk that this unmolten material will rest on the tuyere nose and will cause the tuyere to tip. This can be observed through the peepsight where an oval opening of the tuyere is seen rather than a round one, and has been caused by the tuyere being drawn into the furnace by the heavy weight of the scab bearing down upon it.
The coke requirement at the wall can also be met using nut coke blended into the ore layer. In this case the nut coke is preferentially available for direct reduction and will preserve the larger, metallurgical coke in the layer structure. Note also, that the direct reduction percentage in the wall area can be higher than estimated above, so that even more coke is required at the wall.

7.6.2 Coke layer thickness

When reaching higher and higher coal injection levels the question arises as to whether a minimum coke layer thickness exists, and what would it be?
The gas ascending the furnace from the tuyeres to the top is distributed through the coke layers, so the coke layers must be present at all elevation of the furnace for this to continue. As the layers are made up of discrete coke particles, the theoretical minimum coke layer thickness translates into a number of coke particles. To produce a path for the gas it is considered that the minimum number of coke particles that should be present in the height of one layer is three. The minimum thickness is therefore three times the mean size of coke in the belly of the blast furnace. Taking for example an average coke size of 50 mm, it would therefore be reasonable to expect that the minimum coke layer thickness in the belly is 15 cm. As the effective ratio of the surfaces of belly to throat is generally around two, the minimum coke layer thickness at the throat should have a minimum of about 30 cm.
In operational practice of furnaces operating at high coal injection levels, the coke layer at the top have reached values as low as 32 cm.

7.7 Ore layer thickness

What is the effect of ore layer thickness on the process? If thicker ore layers are charged, less ore layers are present in the operating furnace and less coke slits are available to distribute the gas. But, especially in conveyor belt fed furnaces, the thicker the ore layer, the more charging capacity is available.
For reduction and melting two effects must be considered, those being the reduction in the granular zone of the furnace and the melting of the layers in the cohesive zone.

7.7.1 Reduction in granular zone

The reduction capacity of gas entering thicker ore layers will be depleted faster and as a consequence, the reduction of ore burden in the granular zone will be poorer.

7.7.2 Softening and Melting

As soon as an ore layer starts to soften and melt, it becomes impermeable for gas. This means that ore layers are heated up at the contact surface between the coke and ore layer. The thicker the ore layer, the longer it will take to melt down completely. Moreover, the melting of the ore layer slows down because there is more oxygen in the ore layer, because of lower rate of pre–reduction (see preceding section). So the thicker the ore layer, the more difficult the melting of the layer (Figure 7.20, next page).

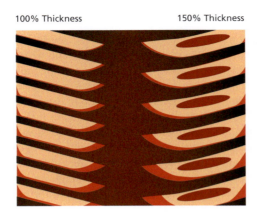

Figure 7.20   Melting of thin and thick ore layers compared

7.7.3 Optimizing ore and coke layer thickness

So, the blast furnace operator wants good permeable coke layers (i.e. thick layers) and good melting ore layers i.e. thin layers. As is often the case in BF operation the best operational results can only be reached with a compromise between these two factors. Generally speaking, from operational observation, the ore layers should not exceed 70–80 cm in the throat of a blast furnace and coke layers should not be smaller than 32 cm. The operational optimization depends on local situations.

Experience has shown that:
– Permeable ore layers can be maintained even when the layers have become quite thick, provided a permeable ore burden is used. For pellet burdens this would require screening of the pellets, and for sinter it would have to be sized to a relatively large diameter (more than 5 mm).
– The minimum coke layer thickness experienced was 14 cm metallurgical coke in the belly.

Conveyor belt fed furnaces tend to work with thicker ore layers. This is caused by the fact that in a conveyor fed furnace the charging capacity increases with increasing layer thickness. In skip–fed furnaces the optimum charging capacity is reached with full skips of coke. In the past the volume of coke was normally the determining factor, so furnaces tended to work with full skips of coke. At high coal injection rates the skip weight is normally the determining factor and thus furnaces now work with full skips of ore.

Another aspect of the optimization of the coke layer thickness has to do with the gas permeability of the coke layer. The coarser the coke is screened in the blast furnace stockhouse, the more permeable the layer is. There are, however, two drawbacks of the coarse (35 mm or more) screening of coke.

Consequence 1: The coarser the coke is screened, the more nut coke or small coke is produced. The nut coke is added to the ore burden layer, increasing the thickness of the ore burden layer and decreasing the size of the coke layer.
Consequence 2: The coarser the coke is screened at the stockhouse, the thicker the formation of a mixed layer at the coke–burden interface.

Optimization depends on local conditions, but high productivity has been reached with a coke screen size in the stockhouse of 25 mm and a nut coke quantity of 25 kg/tHM.

7.7.4 "Ideal" burden distribution

The ideal burden distribution for high productivity and high PCI rates is—according to the authors—as follows:
– An ore free centre,
– Nearly horizontal layers of coke and ore burden,
– Some nut coke in the ore burden in the wall area and
– Coarse coke in the centre.

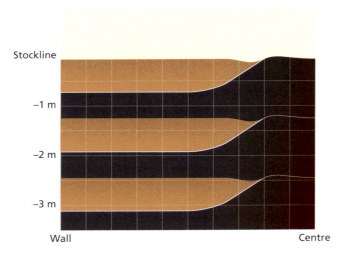

Figure 7.21   "Ideal" burden distribution

Ore free centre
The ore free centre allows the gas to distribute itself through the coke layers from inside to outside. We can consider the coke layers as layers with equal pressure. If the total internal pressure difference in 1.2 bar, the pressure difference over each of the 40 ore layers is about 0.03 bar. The ore free centre typically has a diameter of 1.5 to 2 metres. The ore free centre can be made in a furnace with a bell–less top by discharging 10–15 % of the coke on a very inward chute position. In furnace with a double bell top, formation of an ore free centre is more difficult.

Nearly horizontal layers

Using nearly horizontal layers of coke and ore minimizes the effect of natural deviations of parameters important for the formation of the layers. E.g. wet pellets have a different angle of repose as compared with dry pellets. This does not affect burden distribution if nearly horizontal layers are used. Care should be taken, that there is no inversion of the profile, i.e. a pile in the centre of the furnace. This can be monitored with e.g. a profilemeter.

Nut coke

The gas in the wall area is cooled by the heat losses to the wall. Moreover, in the wall area a relatively large percentage of fine ore burden materials is located and reduction disintegration is strongest (because of slower heating and reduction). For these reactions, reduction and melting of the ore burden in the wall area is most difficult. Nut coke in the wall area helps to reduce reduction gas and heat requirements in the wall area. The nut coke has a lower heat capacity than the ore burden. Moreover, when the ore burden in the wall area starts melting, the nut coke is immediately available for direct reduction. In doing so, it prevents the direct reduction attack on the metallurgical coke.

Coarse coke in the centre

The coke charged in the centre is the least attacked by the solution loss reaction and has the smallest chance to be burnt in front of the tuyeres. Therefore, it is thought that the coke charged in the centre finally constitutes the coke in the hearth. Good permeability of the hearth helps to improve casting and prevents preferential flow of iron along the wall, thus increasing hearth campaign length.

7.8 Erratic burden descent and gas flow

The burden descent sometimes becomes erratic (see Figure 7.3). What happens in the furnace if it hangs and slips? The mechanism of hanging and slipping is illustrated in Figures 7.22–7.24.

First, the furnace hangs because at the cohesive zone, bridges of melting ore burden are formed. "Bridge formation" is the phenomenon, that solid materials can be piled upon each other and will not collapse into a hole: see Figure 7.22 for a bridge formed from marbles.

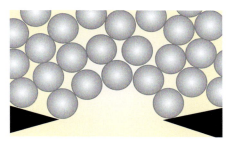

Figure 7.22   Bridge formation illustrated by a theoretical experiment with marbles

Second, while the furnace hangs, the process continues: coke is consumed and ore burden melts. Therefore, voidage arises in or below the cohesive zone.

Third, when this voidage becomes too big, it collapses: the furnace slips. (Figure 7.23). The layer structure is completely disrupted and the gas flow through these layers is impeded. This leads again to areas in the furnace where ore burden is insufficiently reduced and remains in a cohesive state for too long. These areas will form the bridges for next time the furnace hangs. The problem can only be solved by re–establishing the layer structure within the furnace, which means, that the complete content of the furnace has to be refreshed: the furnace has to be operated on reduced blast volume for five to ten hours.

Figure 7.23   Creation of voidage below bridges and consequential collapse

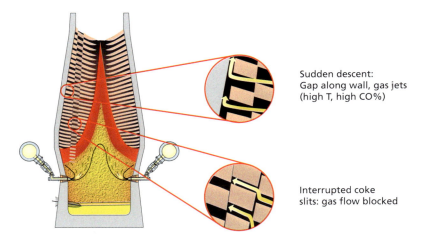

Figure 7.24   Disrupted layer structure and impeded gas flow

After a slip, the layer structure in the furnace is disrupted and therefore the contact between gas and burden is impeded (Figure 7.24). As a consequence, the gas reduction reactions slow down and extra direct reduction will take place in the hearth: the furnace will chill. The process will recover when a normal layer structure is restored. It takes 6–8 hours to refill the furnace on a decreased wind volume.

7.9 Blast furnace instrumentation

An overview of blast furnace instrumentation as discussed in various parts of the text is given in Figure 7.25.

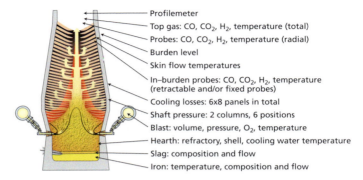

Figure 7.25   Overview of Blast Furnace instrumentation

7.10 Blast furnace daily operational control

In this section the blast furnace daily operational control is discussed. The better the consistency of the blast furnace input, the lower the need for adjustments in the process. Ideally, a good consistency of the input allows the operator to "wait and see". The need for daily operational control is a consequence of the variability of the input and – sometimes– the equipment.

The process must be controlled continuously, which may require changes to be made on a daily or even shift basis. The changes are aimed towards:
– Correct iron and slag composition. The burden and coke are adjusted to get the correct chemical composition of the iron and slag. For the latter especially the basicity of the slag is important because of its effect on hot metal sulphur. Correct iron and slag composition also implies control of thermal level, since the hot metal silicon is correlated with the hot metal temperature. So, there are daily requirements for burden calculations with updated chemical analysis of the burden components and actual burden, and frequent adjustments of the thermal level of the furnace. Adjusting the coke rate or the auxiliary reductant injection through the tuyeres can achieve the latter.
– Stable process control. Burden descent (as measured by the stock rods, Figure 7.1, or pressure taps, Figure 7.26), blast furnace productivity and efficiency are evaluated on the basis of hourly data. Raceway conditions (e.g. flame temperature) are monitored or calculated. The total process overview gives an indication whether or not adjustments are required. Pressure taps indicate whether or not "short circuiting" of gas flow along the wall takes place. In stable periods the layers of coke and ore can be followed when passing the pressure taps.

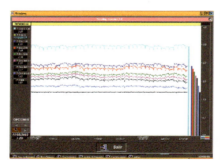 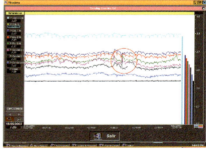

Figure 7.26 Pressure taps indicating the stability of the process, 24hr graphs. The example shows stable (left) and unstable (right) operation, with short–circuiting of gas flow encircled in red. (Courtesy: Siderar, Argentina)

– Gas flow control. The subject of gas flow control is discussed in more detail below. Measurements and data required for daily gas flow control are shown in Figure 7.27. The gas flow through the furnace can be monitored with the help of global top gas composition, top gas composition across the radius, heat losses at the wall and gas flow along the wall. The latter can be measured with the short in–burden probes: the probes measure the temperature about three metres below the burden level up to 50 cm into the burden. If temperatures are low (under 100°C) the burden is not yet dry and more gas flow in wall area is required to increase the drying capacity at the wall.

If the furnace seems in need of an adjustment of the gas flow, a change to the burden distribution can be considered. However, a thorough analysis of the actual situation has to be made. For example, consider the situation whereby high central temperatures are observed. If these high central temperatures are observed together with low heat losses and low gas utilisation, then the central gas flow can be considered to be too high. The appropriate action in this case would be to consider changes to the burden distribution to decrease the central gas flow. If, on the other hand, the high central temperatures are combined with a good gas utilisation and good wall gas flow, then there is no need to change the layers of ore and coke. The appropriate action in this scenario would be to consider working with lower gas volume per tonne HM i.e. with higher oxygen enrichment.

Note also, that the heat losses are very sensitive to the burden descent. Irregular burden descent leads to gaps at the wall and high heat losses. So, if a furnace is showing high heat losses, again, the cause should be investigated in detail before adjusting burden distribution. For example, if a blast furnace is pushed to its production limits and burden descent suffers due to the high pressure difference over the burden, the solution of the high heat losses is to reduce production level (or gas volume) and not to adjust burden distribution.

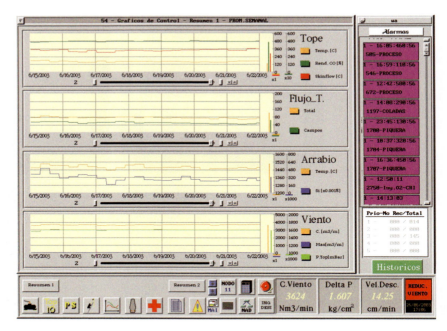

Figure 7.27   Presentation of process data in an operational furnace.
The weekly graph gives an overview from the stability and development of the process. From top downwards: Tope – CO utilisation (%), skin flow temperature (°C) and top temperature (°C); Flujo T – Total heat loss and sum of fields (GJ/hr); Arrabio – Hot metal temperature (°C) and silicon (%); Viento – Blast volume (Nm³/min) and top pressure (bar)

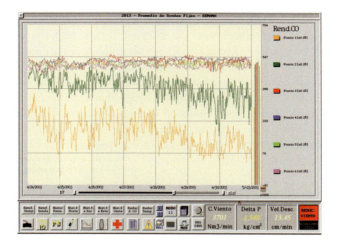

Figure 7.28   Example of gas flow control.
The radial gas distribution is measured with above burden probes, expressed as CO utilisation (7 day graph). The decreasing gas utilisation in the centre of the furnace (point 1 and 2, yellow and dark green) shows increased central working.

VIII *Blast Furnace Productivity and Efficiency*

The production rate of a blast furnace is directly related to the amount of coke used in front of the tuyeres in a stable situation. This is due to every charge of coke at the top of the furnace bringing with it an amount of ore burden materials. In a stable situation the hot metal is produced as soon as the coke is consumed. The productivity of a blast furnace increases as less reductant is used per tonne hot metal. In the present chapter the basics behind blast furnace productivity, the chemical reactions and efficiency are discussed (see also Hartig et al, 2000).

8.1 The raceway

8.1.1 Production rate

In the raceway hot gas is formed which melts the burden material and is used to drive the chemical reactions in the furnace. Given a certain amount of coke and coal used per tonne hot metal, the production rate of a blast furnace is determined by the amount of oxygen blown through the tuyeres. The more oxygen that is blown into the furnace, the more coke and coal are consumed and form carbon monoxide (CO), and the higher the production rate becomes. In addition, the lower the reductant requirement per tonne of hot metal (tHM), the higher the production rate. A quantitative example is indicated below. Coke (and coal) are not only gasified in front of the tuyeres, but are also used for carburisation of iron (hot metal contains 4.5% C) and for direct reduction reactions (section 8.2). The coke rate is expressed as standard coke, i.e. coke with a carbon content of 87.5 %.

In an operating blast furnace the use of the reductants can be as follows:

	Input (kg/tHM)	Replacement ratio	Input, as standard coke (kg/tHM)
Coke	300	n/a	300
Coal	200	0.85	170
Total			470

	Use, as standard coke (kg/tHM)
Total input	470
Carburisation	50
Direct Reduction	120
Gasified in front of tuyeres	300
Of which coal	170
And coke	130

Table 8.1   Reductants in a blast furnace, typical example

The 300 kg/tHM standard coke which is used in front of the tuyeres consists of 170 kg/tHM coke equivalent injected as coal and so per tonne hot metal, 130 kg coke (300–170 kg) is gasified at the tuyeres. Note the issue of efficiency: if the same amount of oxygen is blown into the furnace, thus maintaining same blast volume and blast conditions, while the reductant rate is 10 kg/tHM lower, the production rate will increase. At a 10 kg/tHM lower reductant rate the production will increase by 3 % (300/290–100)! Conversely, if extra coal is put on the furnace for thermal control, the production rate will decrease if blast conditions are maintained. This is a simplified approach. Secondary effects, like the effect on gas flow throughput, the effect on flame temperature and the oxygen content of the coal, have been neglected.

8.1.2   Bosh gas composition

The heat of the blast and the heat generated by the reactions of coke (and coal/auxiliary reductants) in the raceway are used to melt the burden. The heat available to melt the burden depends on the amount of gas produced and on the flame temperature, known as the "raceway adiabatic flame temperature" (RAFT).

The amount and composition of the raceway gas can be calculated using the following reactions that take place in the raceway:

$$2\,C + O_2 \leftrightarrow 2\,CO$$
$$H_2O + C \rightarrow CO + H_2$$

In and directly after the raceway all oxygen is converted to carbon monoxide and all water is converted to hydrogen and carbon monoxide.

Consider the following example; the blast furnace in section 2.3 has a blast volume of 6,500 m³ STP with 25.6 % oxygen. Ignoring the effects of moisture in the blast and the coal injection, what would be the raceway gas volume and composition?

Blast into the furnace (per minute):
- Nitrogen: 4836 m³ STP/min ((1–0.256)x6500)
- Oxygen: 1664 m³ STP/min (0.256x6500)

The oxygen generates two molecules of CO for every O_2 molecule, so the gas volume is 8164 m³ STP/min (4836+2x1664). The gas consists of 59 % nitrogen (4836/8164) and 41% CO (2x1664/8164).
The calculation can be extended to include the moisture in the blast and the injection of coal (or other reductants). This is done in section 6.4.

8.1.3 Raceway flame temperature

The flame temperature in the raceway is the temperature that the raceway gas reaches as soon as all carbon, oxygen and water have been converted to CO and H_2. The flame temperature is a theoretical concept, since not all reactions are completed in the raceway. From a theoretical point of view it should be calculated from a heat balance calculation over the raceway. For practical purposes linear formulas have been derived (see e.g. Table 8.2).

		Metric Units
RAFT =	1489 + 0.82xBT – 5.705xBM + 52.778x(OE) – 18.1xCoal/WCx100 – 43.01xOil/WCx100 – 27.9xTar/WCx100 – 50.66xNG/WCx100	
Where	BT	Blast Temperature in °C
	BM	Blast Moisture in gr/m³ STP dry blast
	OE	Oxygen enrichment (% O_2 – 21)
	Oil	Dry oil injection rate in kg/tHM
	Tar	Dry tar injection rate in kg/tHM
	Coal	Dry coal injection rate in kg/tHM
	NG	Natural gas injection rate in kg/tHM
	WC	Wind consumption in m³/tHM

Table 8.2   RAFT Calculation (source: AIST)

Flame temperature is normally in the range of 2000 to 2300°C and is influenced by the raceway conditions. The flame temperature increases if:
- Hot blast temperature increases.
- Oxygen percentage in blast increases.

The flame temperature decreases, if:
- Moisture increases in the blast.
- Reductant injection rate increases, since cold reductants are gasified instead of hot coke. The precise effect depends also on auxiliary reductant composition.

Table 8.3 gives some basic rules with respect to flame temperature effects.

	Unit	Change	Flame temp. (°C)	Top temp. (°C)
Blast temp.	°C	+ 100	+ 65	– 15
Coal	kg/t	+ 10	– 30	+ 9
Oxygen	%	+ 1	+ 45	– 15
Moisture	g/m³ STP	+ 10	– 50	+ 9

Table 8.3   Flame temperature effects, rules of thumb (calculated)

The top gas temperature is governed by the amount of gas needed in the process; the less gas is used, the lower the top gas temperature and vice versa. Less gas per ton hot metal results in less gas for heating and drying the burden.

8.2 Carbon and iron oxides

In the preceding section the formation of gas in the raceway has been described. What happens with the gas when it ascends through the furnace and cools down? First consider what happens with the carbon monoxide.

Carbon can give two types of oxides:
- $C + ½ O_2 \rightarrow CO$ + heat (111 kJ/mole)
This reaction takes place in the blast furnace
- $C + O_2 \rightarrow CO_2$ + heat (394 kJ/mole)
This reaction does not take place in the raceway and is more typical in an area such as a power plant.

Note that in the second step much more heat is generated than in the first step, therefore, it is worthwhile to convert CO to CO_2 as much as possible in the process. The ratio $CO_2/(CO+CO_2)$ is called the gas utilisation or gas efficiency and is used extensively in blast furnace operation.

In Figure 8.1, the equilibrium $CO \leftrightarrow C + CO_2$ is presented for various temperatures. The line indicates the equilibrium of the "Boudouard" reactions. At temperatures above 1,100°C all CO_2 is converted to CO, if in contact with coke. Therefore, at the high temperatures in the bosh and melting zone of the blast furnace only carbon monoxide is present. At temperatures below 500 °C all CO has the tendency to decompose into $C+CO_2$. The carbon formed in this way is very fine and is called "Boudouard" carbon.

In operational practice the carbon monoxide decomposition can be observed in refractory material, where there is a CO containing atmosphere in the correct temperature region. This generally is a very slow process.

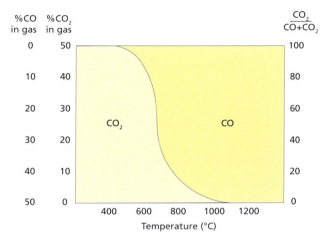

Figure 8.1   Boudouard reaction: the drawn line indicates equilibrium

8.2.1   Direct reduction of iron oxides

As the hot reducing gases produced in the raceway ascend through the lower furnace, they transfer heat to the ore burden to the extent that it becomes molten at the lower levels of the melting zone. They also remove oxygen from the iron oxides, i.e. they reduce the iron oxides, which contain approximately one oxygen for every two iron atoms.

The CO_2 produced from the reaction immediately reacts with the carbon in the coke to produce CO. The total reaction is known as direct reduction, because carbon is directly consumed.
The reactions can be indicated as below:

$$2\ FeO_{0.5} + CO \rightarrow 2\ Fe + CO_2$$
$$+\quad CO_2 + C \rightarrow 2\ C$$
$$\text{Total}\ 2\ FeO_{0.5} + C \rightarrow 2\ Fe + CO \quad \textit{(consumes 155 kJ/kmol FeO)}$$

The direct reduction reaction requires an enormous amount of heat, which is provided by the heat contained in the hot raceway gas.

The direct reduction reaction is very important for understanding the process. In a modern blast furnace the direct reduction removes about a third of the oxygen from the burden, leaving the remaining two–thirds to be removed by the gas reduction reaction. The amount of oxygen to be removed at high temperatures, as soon as the burden starts to melt, is very much dependent on the efficiency of the reduction processes in the shaft. See section 8.2.2.

Note the following important observations:
- Direct reduction uses carbon (coke) and generates extra CO gas.
- Direct reduction costs a lot of energy.

In operational practice the direct reduction can be monitored. In many blast furnaces the direct reduction rate (the percentage of the oxygen removed from the burden by direct reduction) or the solution loss (the amount of coke used for the reaction) are calculated on line. Experienced operators are well aware that as soon as the direct reduction rate or the solution loss increases, the blast furnace starts to descend faster, the cohesive zone will come down as the coke below it is consumed. And the furnace will chill. When properly observed, chilling can be prevented, for example by using extra coal injection.

8.2.2 Direct reduction of accompanying elements

In addition to the direct reduction of iron (typically from $FeO_{0.5}$) some other materials are also directly reduced in the high temperature area of the furnace. The amount of coke used for this direct reduction reactions is indicated in the table below. This can be calculated from the chemical composition and the atomic weights, considering that the amount of oxygen removed reacts with the carbon in the coke. The 121.9 kg coke for direct reduction corresponds with an additional 199 m³ STP of CO gas.

Material	Reduced to	Typically in hot metal (%)	Coke used (kg/tHM)
$FeO_{0.5}$	Fe	94.50	116.1
SiO_2	Si	0.40	3.9
MnO	Mn	0.30	0.7
TiO_2	Ti	0.05	0.3
P_2O_5	P	0.07	0.8
Total coke used for direct reduction			121.9

Table 8.4   Coke consumption for direct reduction, typical example

8.2.3 Gas reduction or "indirect" reduction

As soon as temperatures of the gas reduce, the CO_2 becomes stable and reduction reactions can take place, such as (see Figure 8.2):
- For Haematite:
  $6\ Fe_2O_3 + 2\ CO \rightarrow 4\ Fe_3O_4 + 2\ CO_2$ (generates 53 kK/kmol)
- For Magnetite:
  $4\ Fe_3O_4 + 4\ CO \rightarrow 12\ FeO + 4\ CO_2$ (consumes 36 kK/kmol)
- For Wustite:
  $6\ FeO + 3\ CO \rightarrow 6\ FeO_{0.5} + 3\ CO_2$ (generates 17 kK/kmol)

The reduction is called "gas reduction" because the oxygen is removed from the burden materials with CO gas. H_2 reacts in a similar way. In literature it is also often called "indirect" reduction, since carbon is only indirectly involved in this reaction. The reduction of the $FeO_{0.5}$ takes place via the direct reduction.

Following the burden descent from the stockline, the reduction from haematite to magnetite starts around 500°C. The reduction from magnetite to wustite takes place in the temperature zone from 600 to 900°C, while the reduction from wustite to iron takes place in the temperature region between 900 and 1,100°C. At the start of melting (1,100 to 1150°C) $FeO_{0.5}$ is normally reached. Here FeO is used as a symbol for wustite, however the most stable composition is $FeO_{0.95}$. The reactions are shown in Figure 8.2.

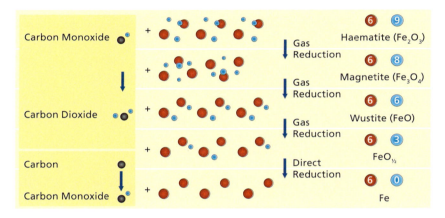

Figure 8.2 Overview of the reduction of iron oxides
(Black dots are carbons atoms, blue dots hydrogen atoms, red dots iron atoms)

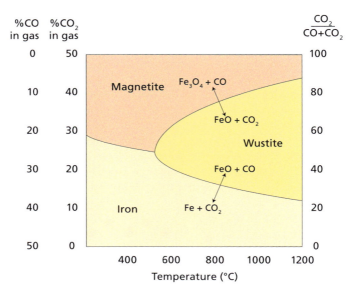

Figure 8.3   Schematic presentation of the relation between temperatures, CO/CO_2 gas composition and iron oxides. The drawn lines indicate equilibrium.

The equilibrium between the various iron oxides and the gas is shown in Figure 8.3. The figure shows at what level of temperatures and gas compositions further gas reduction of the burden is no longer possible. The reduction of wustite to iron requires gas with a relatively high percentage CO. Gas utilisation for reduction of wustite should be below 30 %. If CO_2 is higher, wustite is no longer converted to iron.

The progress of the reduction reactions in a blast furnace can be detected in two different ways:
– Burden: from quenched furnaces an overview of the progress of the reduction can be derived. An example is shown in Figure 8.4
– Gas: by sending gas sampling devices down into the furnace, the progress of temperature/gas composition can be derived. Figure 8.5 shows typical results from a gas sampling exercise. The data can be depicted in the graph of the equilibrium between gas and iron oxides. The gas normally shows a "thermal reserve zone", that is, a zone in which the temperature does not change rapidly as well as a "chemical reserve zone", a zone in which the chemical composition of the gas does not change. The thermal reserve zone decreases and can disappear when the furnace is pushed to high productivities.

Blast Furnace Productivity and Efficiency

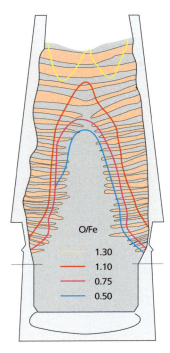

Figure 8.4 Reduction progress in a quenched furnace (Hirohata, after Omori, 1987, p. 8)

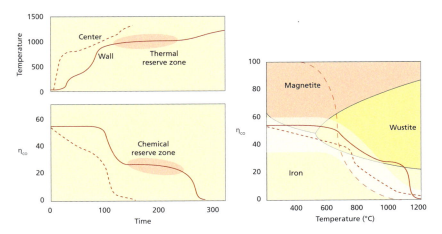

Figure 8.5 Gas composition in operating furnace. CO, CO_2, H_2 and temperature were measured with descending probes (Chaigneau et al, 2001). Typical measurements from various furnaces are shaded. After McMaster, 2002.

8.2.4 Gas reduction and direct reduction

The direct reduction and gas reduction reaction combine to create a very efficient process. Suppose that all oxygen is removed by direct reduction. Then, the following reaction takes place:

$$Fe_2O_3 + 3\,C \rightarrow 2\,Fe + 3\,CO$$

Iron contains about 945 kg Fe per tonne hot metal. Coke contains about 87.5% carbon. Atomic weights of Fe and C are 55.6 and 12 respectively. A tonne of iron contains 17 kmole (945/55.6). For every atom of iron we need 1.5 atoms of carbon, so the carbon requirement is 25.5 kmole (1.5x17), which is 306 kg carbon (25.5x12). In addition, about 45 kg carbon is dissolved in iron. In total, 351 kg carbon is used per tonne hot metal, which corresponds to 401 kg of coke. This is a very low equivalent coke rate and a blast furnace will not work, because the heat generated in this reaction is too low.

Now consider that all reduction reactions are done via the gas reduction, what coke rate is required in this situation? It is assumed that coke combustion generates the CO required. The reaction is:

$$3\,FeO + 3\,CO \rightarrow 3\,Fe + 3\,CO_2$$

We only consider the reduction of wustite since the resulting gas is powerful enough to reduce magnetite and haematite. We know from the above (Figure 8.3) that for gas reduction the maximum gas utilisation is 30%. To get 30% gas utilisation more CO is needed and the reaction becomes:

$$3\,FeO + 10\,CO \rightarrow 3\,Fe + 3\,CO_2 + 7\,CO \quad \textit{(gas utilisation: 3/(3+7) = 30\%)}$$

So the coke requirement is calculated as above: every tonne iron contains 17 kmole.

There is a need of 10 atoms carbon per 3 atoms of Fe. So the carbon requirement is 57 kmole (10/3x 17), which corresponds to 684 kg carbon (57x12). Again, the extra 45 kg carbon in iron has to be added giving a carbon rate of 729 kg/t and a coke rate of 833 kg coke per tonne hot metal (729/0.875). This reaction has a poor coke rate and a high heat excess.

The conclusion of the considerations above is, that the counter–current character of the blast furnace works efficiently to reduce the reductant rate by combining direct and gas reduction reaction. Approximately 60–70% of the oxygen is removed by gas and the remaining oxygen is removed by direct reduction.

8.2.5 Hydrogen

Hydrogen is formed from moisture in the blast and injectants in the raceway. Hydrogen can act as a reducing agent to remove oxygen and form water. The reaction is comparable with that for carbon monoxide:

$$H_2 + FeO \rightarrow Fe + H_2O$$

The major differences with the reactions for hydrogen and carbon monoxide are as follows:
- Figure 8.6 shows the equilibrium of the iron oxides and hydrogen. Hydrogen is more effective for the reduction at temperatures above 900 °C. From measurements in the blast furnace it has been derived, that hydrogen reactions are already nearly completed at this temperature.
- Hydrogen utilisation as measured from the top gas is normally around 40 % while CO utilisation is close to 50 %. At the FeO level (900 °C) hydrogen is utilized for 35 %, which means that it is already close to its final utilization of 40 %.
- Hydrogen is less effective a reductant at lower temperatures, becuase it generates less heat when reducing iron oxides.

At high temperatures H2O that is formed in the furnace reacts with coke according to the water–gas–shift reaction:
H_2O *(steam)* $+ \; C \; \leftrightarrow \; H_2 \; + \; CO$ *(consumes 124 kJ/mole)*
This reaction consumes a lot of heat. At higher temperatures (over 1000 °C) the reaction proceeds rapidly to the right hand side. This reaction is particularly manifest when a furnace is blown down: water vapour is in contact with CO_2 rich, hot top gas (see also section 11.5).

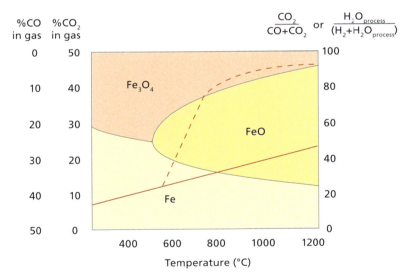

Figure 8.6   Equilibrium iron oxides with hydrogen and carbonmonoxide

Note that the hydrogen utilisation cannot be measured directly. The H_2O formed in the process cannot be discriminated from the water put in the furnace with coke and burden moisture. The hydrogen utilisation of the top gas is defined as $H_2O/(H_2+H_2O_{process})$. The $H_2+H_2O_{process}$ can be derived from the input, the hydrogen leaving the furnace can be measured with the gas analysis.

When working at high hydrogen input (via moisture, natural gas, coal), the competition between the reduction reactions will lead to lower top gas CO_2 utilisation. The simple reasoning is, that H_2 competes with CO. All oxygen taken by H_2 is not taken by CO_2 and thus CO increases and CO_2 decreases. 1 % extra H_2 in topgas will lead to 0.6 % extra $H_2O_{process}$ in top gas and thus to a 0.6 % lower CO_2 and a 0.6 % higher CO percentage. 1 % extra topgas hydrogen leads to a decrease in topgas CO–utilisation of 1.3 %, e.g from 49 % to 47.7 %. If a more advanced model is used and the efficiency of the furnace is kept constant at the FeO level, a 1% increase in topgas hydrogen leads to a decrease of 0.8 % in topgas CO–utilisation.

8.3 Temperature profile

The temperature profile and the chemical reactions in a blast furnace are closely related. It is summarised in Figure 8.7. The reduction of the oxides to wustite takes place at temperatures between 800 and 900 °C. Thereafter, in the temperature range of 900 to 1100 °C, the wustite can be further reduced indirectly without interference from the Boudouard reaction. This chemical preparation zone can take up to 50 to 60 % of the height of the furnace and has a relatively constant temperature. This region is called the thermal reserve zone.

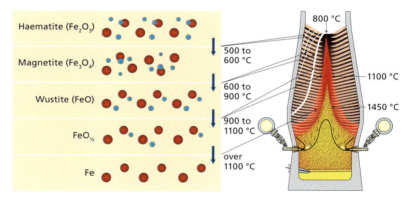

Figure 8.7   *Progress of the reduction reactions and temperature of the burden*

8.4 What happens with the gas in the burden?

In the preceding section the temperature profile in the blast furnace has been shown. In this section the gas in the furnace will be dealt with in more detail.

Step 1   Wind is blown into the tuyeres along with coal and moisture. All these components react to form carbon monoxide (CO), hydrogen and nitrogen. So, the conditions at the end of the raceway are a high temperature of 2000 to 2200 °C and CO, H_2 and N_2 in gaseous form.

Step 2   The gas ascends in the furnace and cools down to 1100 °C. The direct reduction reactions take place generating additional CO gas. When reaching 1100 °C the gas leaves the cohesive zone and enters into the furnace stack filled with granular materials. At temperatures over 1100 °C gas reduction is very limited as the CO_2 formed by direct reduction reacts instantaneously with coke to return to CO, a reaction which is thermodynamically equivalent to direct reduction.

Step 3   The gas ascends further and its temperature decreases from 1100 °C to 900 °C. In this temperature range the hydrogen is very effective and about 35 % of the hydrogen picks up an oxygen from the ore burden. About 24 % of the carbon monoxide does the same.

Step 4   The gas ascends further reaching an area of 500 to 600 °C. At this temperature the ore burden has the composition of magnetite, Fe_3O_4.

Step 5   The gas cools down further to the temperature at which it will leave the top (110 to 150 °C). In this area the carbon monoxide is utilized further and removes more oxygen from the ore burden.

In terms of gas volume, once the temperature of the gas has dropped below 1100 °C, the total gas volume in m³ STP remains the same, and only the composition of the gas changes, as shown in figure 8.8.

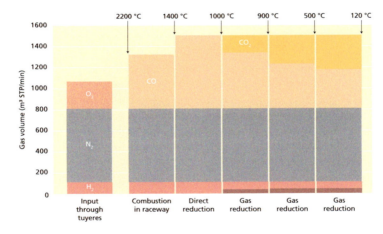

Figure 8.8   How top gas is formed from wind

It is clear from figure 8.8, that the major part of the gas through the furnace consists of nitrogen. Nitrogen is chemically inert and delivers only its heat from the hot blast to the burden. During its eight to twelve second journey through the furnace it cools down from the blast temperature to the top gas temperature.

8.5 Oxygen and productivity

The productivity of blast furnaces vary enormously and in this section the question of what is the optimum productivity a furnace can reach is tackled. The first point to consider is that the furnace is a gas reactor, so the more blast you blow into the furnace, the more it produces. The furnace can be driven to a maximum ΔP, but when going to a ΔP over the burden above this maximum, the burden descent will deteriorate and the process will slow down.

Therefore the first limit is maximum ΔP, or, with constant top pressure, maximum blast pressure. The 'allowable' maximum ΔP depends on the furnace and the burden and is a specific value for each furnace. Large 14 metre hearth diameter furnaces can work up to levels of a maximum ΔP of 1.95 bar. Note that this is an instantaneous maximum. Systems which automatically lower the blast volume when reaching the maximum ΔP help to prevent hanging, slips and process upsets.
Secondly, consider how the furnace works: the 900 °C isotherm is also the plane where the ratio Fe/O (on atomic basis) is 1. So, between the 900 °C isotherm and the tuyeres the gas has to be able to melt the burden, to reduce the remaining oxides and execute the other direct reduction reactions. So, the heat balance over the lower part of the furnace has to be closed. At 900 °C the remaining gas and heat is used for the heating up the burden and coke from ambient temperature to 900 °C and reduction of the oxides. So the net effect translates to a top gas temperature. If top gas temperature is high, well above 110 °C, the gas volume in the upper part of the furnace increases as gas expands at higher temperatures. Therefore, the productivity is highest if the top gas temperature is at its minimum. The minimum we recommend is to stay above the dewpoint so that all moisture is driven off. This translates to a narrow band slightly above 100 °C.

Since the top gas temperature decreases when using higher oxygen, the maximum productivity in a furnace (with a given burden) is reached when;
– Top pressure is at the maximum value.
– Blast volume is set so that the furnace is operated to the maximum ΔP.
– Fuel injection (coal, gas, oil) is at the maximum the furnace accepts.
– The top gas temperature is controlled to slightly above 100 °C with oxygen injection.

The quality of the burden affects the productivity by means of the coke rate: the lower the coke rate per tonne hot metal, the more hot metal can be produced with the same amount of blast. On average for every 3–3.5 kg/tHM decrease in coke rate (or coal rate) the productivity increases by 1 %.

8.6 Use of metallic iron

Metallic iron can be used to boost the productivity of the furnace. The metallic units can be scrap, but Hot Briquetted Iron (HBI) can also be used. As a rule of thumb: about 250 kg per tonne hot metal of coke (or coal) is required to generate the heat for melting metallic iron and to provide the carbon that dissolves in the hot metal. When charging 10% metallic iron units, fuel rate decreases from approximately 500 kg/tHM to 475 kg/t and productivity increases by (500–475)/3.5=7 %.

8.7 How iron ore melts

This section deals with the subject of hot metal and slag formation in and around the cohesive zone of the blast furnace.

8.7.1 Ferrous burden

Ferrous burden is the collective term used to describe the iron–containing materials that are charged to the furnace, namely, sinter, pellets and lump ore. The melting properties of these materials depend on the local chemical slag composition. Lump ore has its natural slag composition as it is found, gangue consists of mainly acid components like SiO_2 and Al_2O_3. Pellets and sinter have an artificial composition with components added to the natural iron ores, such as limestone ($CaCO_3$), dolomite ($MgCO_3.CaCO_3$), olivine ($2MgO$, SiO_2) and others. Sinter has a basicity, $CaO/SiO_2 > 1.6$ and, can even be as high as 2.8 or higher. Pellets have a wide variety of chemical compositions, especially acid pellets ($CaO/SiO_2 < 0.2$) or fluxed pellets ($CaO/SiO_2 > 0.8$).

The chemical composition of the materials is not only based on the design of the optimum properties of the final slag, with respect to fluidity and desulphurizing properties, but also on the design of the metallurgical properties of the sinter and the pellets. Optimal metallurgical properties means that the materials should have good reduction–disintegration properties and melting temperatures as high as possible. The reason for these requirements is defined by the nature of the blast furnace process, that being a gas–reduction process. If material falls apart in small particles, the gas flow through the ore layer is impeded and the normal reduction process is limited. In addition, materials which start to melt form an impermeable layer and will also affect the reduction progress.
Note that the efficiency of a blast furnace is largely determined by the gas reduction process, and the amount of oxygen bound on the iron, which is removed by gas (CO and H_2).

8.7.2 Reduction from haematite to magnetite and reduction–disintegration

The reduction process starts at temperatures of about 500 °C in the atmosphere of a reducing gas, that is, the blast furnace top gas. The reduction of haematite (Fe_2O_3, Fe/O = 1.5) to magnetite (Fe_3O_4, Fe/O 1.33) takes place rather easily and generates a small amount of heat. In haematite 6 atoms of iron are bound to 9 atoms of oxygen, which changes to 8 atoms of oxygen upon transition to magnetite. The extra oxygen is bound to the CO gas, forming CO_2.

The first step in the reduction process has a profound effect on the properties of the ferrous burden. The crystal structure where 6 iron atoms and 9 oxygen atoms were happily conjoined is forced to change to 6 iron atoms on 8 oxygen atoms. The crystal structure changes and this leads to stress within the particles and the particles can fall apart. This is called reduction disintegration, and is represented by the Reduction Disintegration Index (RDI) or, more commonly in the USA by Low Temperature Breakdown (LTB). Pellets are not very prone to reduction disintegration, as pellets have about 30 % voidage in the structure, which can take care of local expansion. Moreover, pellets form a solid shell so they retain their round shape and do not impede the local permeability for gas.

Some lump ores have a very tight structure and are difficult to reduce, with the reduction starting on the outside of the particle. These lump ores will have reasonable RDI values, however, if a lump ore has a relatively open structure, which is easily permeable for gas, then the RDI will be poor. Lump ored with this characteristic are not suitable for direct use in the blast furnace.

Sinter, on a micro–scale has a relatively tight structure with limited possibilities for local expansion. Therefore, sinter has inherently poor RDI unless measures are taken to improve it. The RDI can be improved by impeding the formation of the secondary haematite on the sinter strand. Secondary haematite is the material which is reoxidized on the sinter strand, from magnetite back to haematite. This takes place when sinter is cooled with air. It is these secondary haematites that are very prone to reduction disintegration in the blast furnace. The reduction disintegration stops when all haematite is reduced to magnetite.

8.7.3 Gas reduction of magnetite

The magnetite (Fe_3O_4) is further reduced by gas (CO and H_2) to wustite ($FeO_{1.05}$). At around 900 °C equilibrium is reached between the reducing power of the gas and the composition of the iron oxides, that is the FeO level of one atom of oxygen per atom of iron. In this area the temperature is relatively constant (thermal reserve zone), as is the chemical composition of the gas (chemical reserve zone). When blast furnaces are operated at very high productivities, these reserve zone becomes smaller and are ultimately eliminated.

At temperatures around 900 °C the temperature of the coke is still too low to react with the CO_2 gas. The coke reactivity reaction ($CO_2 + C \rightarrow 2\ CO$) starts around 1050 °C. Therefore, all reduction is taking place by means of gas reduction: ($Fe_2O_3 + CO \rightarrow 2\ FeO + CO_2$), and in this temperature range also for a small part by ($Fe_2O_3 + H_2 \rightarrow 2\ FeO + H_2O$). The gas reduction continues to a gas temperature above 1000 °C and a reduction of iron oxide to a level of $FeO_{0.45}$. The higher the temperature, the more H_2 contributes to the gas reduction. The gas reduction continues to rise until the temperature has risen to that where the coke reactivity reaction begins. If material starts to soften and melt (around 1100 °C) the direct reduction reaction ($FeO + C \rightarrow Fe + CO$) will take place.

8.7.4 Melting

Melting starts at local chemical compositions with the lowest melting temperatures. This is where there are high local concentrations of SiO_2 and FeO. Internal migration of atoms will cause larger and larger parts of the particles to soften. The first internal 'melts' of material will form at around 1100 °C and will consist of 60 % gangue and 40 % FeO. In the case of fluxed pellets the first melts will form at around 1150°C with a gangue/FeO ratio of 70/30 %. If the basicity increases further, the starting temperature of melt formation increases to close to 1200 °C, where even less FeO is required. However at the basicity of superfluxed sinter, the formation of melts require again high FeO%, up to 50–60 %. This explains why reduction melting tests of superfluxed sinter generally show a relative large part of residual material, that cannot be melted even at temperatures up to 1530 °C.

When gangue starts to melt, it will come into contact with the slag components of other parts of the ore burden and the slag composition will be averaged. This happens at high FeO concentrations.

Note that a sponge iron skull around a particle has a much higher melting temperature than hot metal. The sponge iron does not yet contain carbon and its melting temperature comes closer to the 1535 °C of the elemental iron temperature, rather than the 1147 °C of iron with 4.2 % carbon content.
In summary, the first melts that are formed in the blast furnace come from acid slag components mixed with iron oxides ($FeO_{0.45}$) and iron. As soon as melts are formed the ore bed collapses. The order of events are firstly that the lump ore structure collapses, due to the acidic gangue, next the collapse of sinter structure followed by the collapse of the pellet structure. As soon as the layers are collapsed, the permeability for gas decreases. It is estimated that permeability for gas disappears more or less completely between 1200–1350 °C. In that situation the layers of cohesive material are only heated with gas flowing along its surface. Reduction by hydrogen plays a special role in this situation. Since hydrogen can easily diffuse into a more solid structure, the hydrogen reduction continues after CO reduction has stopped.

When the melts are heated further and start to drip, the melt consists of a blend of the gangue, FeO and finely dispersed iron, which has not been separated from the melt. The first process in the 'primary' melt is that the gangue loses its FeO. As soon as the FeO is removed and the primary melt flows over coke, the iron starts to dissolve carbon from the coke, which lowers the melting temperature rapidly. This has the affect of making the iron much more liquid when flowing over coke. The carbon of the coke diffuses into or is taken up by the metallic Fe, allowing the iron droplets to separate from the primary melt. After this process has taken place, the iron starts to increase in silicon content, which comes from the SiO gas that was created in the raceway flame.

It is thought that the iron diffuses out of the primary melts and reaches the hearth faster than the slag components. When blowpipes have been filled with bosh slag (primary slag) finely distributed iron has never been observed within the slag. This is attributed to the improved fluidity of the iron due to the carbon dissolution from the coke into the melt, dramatically lowering the melting temperature. This means that the iron droplets will pass through a layer of slag. As long as the slag contains FeO, the silicon in the hot metal will be oxidized back to SiO_2 and the FeO in the slag reduced to Fe. As a consequence, the hot metal formed and dripping down in the centre of the furnace will have high silicon and the hot metal formed at the wall will have low silicon. The final silicon level observed during a cast is a blend of these two 'hot' and 'cold' components.

The formation of the final composition of hot metal and slag is a stepwise process, which is illustrated in Figure 8.9.

Blast Furnace Productivity and Efficiency

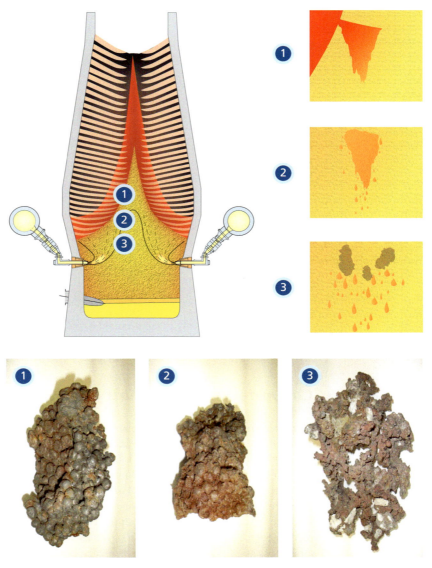

Figure 8.9 Melting of Iron Ore:
1—Material dripping down at 1400 °C, high FeO content (30–40 %), reduced iron contains no carbon and melts at temperatures over 1500 °C
2—Iron separates from the melt at 1500 °C, flowing over coke, carbon dissolves (decrease in melting point to 1170 °C), FeO content low
3—Iron droplets have separated from slag, silicon content increases
(Photographs courtesy J. Ricketts, ArcelorMittal)

In comparison: in a blast furnace, the process of oxygen steelmaking is reversed. With oxygen steelmaking, the elements removed from the hot metal by blowing oxygen are first silicon and manganese, which are oxidised, then carbon is burnt and finally iron starts to be re–oxidised. In the blast furnace, the opposite takes place as is illustrated in figure 8.10.

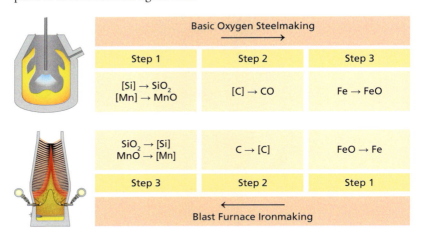

Figure 8.10 The basic oxygen furnace and blast furnace as counterparts
(Rectangular brackets indicate that the element is dissolved in hot metal)

8.8 Circumferential symmetry and direct reduction

High performance operation of a blast furnace requires that the complete circumference of the furnace contributes equally to the process. A furnace can be divided into sectors in which every tuyere forms one sector. See Figure 8.12 for an example. If all sectors do not contribute equally to the process, asymmetry in the melting zone will arise, as shown in Figure 8.11. Local heat shortages will drive the melting zone downwards in certain sectors and upwards in other sectors. This can result in an increase in direct reduction in some sectors. Increasing the thermal level of the entire furnace affecting its overall efficiency can only compensate for the effect and not resolve it.

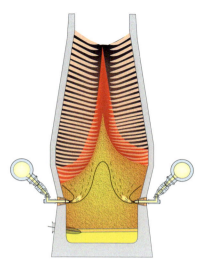

Figure 8.11   Asymmetric melting zone

Asymmetry in the process can arise from various sources:
- By asymmetry of the charging. With a bell–less top this can be prevented by alternating the coke and ore top bins and by changing the rotational direction of the chute. With a double bell system it is possible to alternate the last skip in a dump. Note that the changes have to be made on a time scale smaller than the blast furnace process i.e. more frequent than every six hours.
- Blast distribution: if the blast speed is too low (under 100 m/s), tuyeres will not efficiently function as blast distributors. This can be observed especially at the tuyeres opposite the inlet between hot blast main and bustle main. Blast distribution can also be effected by plugged tuyeres (above a taphole or refractory hot spots) and slag deposits in the tuyere.
- Worn refractory or armouring plates at the top of the furnace.
- From uneven coal injection. Especially tuyeres without PCI (section 5.6).
- Deviation of furnace centre line from vertical line. This is especially a concern in older furnaces.

Measures to correct for deviations of circumferential symmetry are available, such as removing PCI injection from specific tuyeres. However, it is preferred to eliminate the causes of the circumferential asymmetry instead of correcting for it.

Asymmetry in the gas flow can be derived from the radial heat loss distribution. In the figure below, the heat losses are measured in eight segments of the furnace over four vertical sections. Extended asymmetry can be investigated with the help of this type of data and graphs.

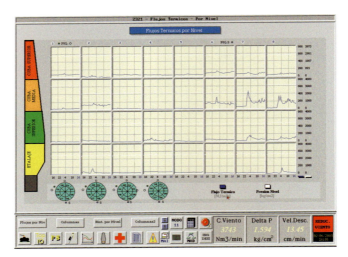

Figure 8.12   *24 hrs heat loss distribution (blue). Note a slight process asymmetry. One day graph of eight sections, four levels.*

IX Hot Metal and Slag

Typical hot metal and slag compositions are given in Table 9.1. Hot metal leaves the furnace with a temperature typically in the range between 1480 and 1520 °C.

Hot metal		Typical	Slag	Typical	Range
Iron	Fe	94.5 %	CaO	40 %	34–42 %
Carbon	C	4.5 %	MgO	10 %	6–12 %
Silicon	Si	0.40 %	SiO_2	36 %	28–38 %
			Al_2O_3	10 %	8–20 %
Manganese	Mn	0.30 %			
Sulphur	S	0.03 %	Sum	96 %	
Phosphorous	P	0.07 %	Sulphur	1 %	

Table 9.1   Typical hot metal and slag composition

9.1 Hot metal and the steel plant

Hot metal is used for the production of steel. In a steel plant the hot metal is refined so that the (chemical) composition can be adjusted to the metallurgical requirements. The refining process is usually achieved in two steps:
– Removal of sulphur from the hot metal by means of desulphurisation. In most cases the sulphur is removed with carbide and lime (stone) or magnesium, according to:
$$2\,CaO + 2\,[S] + CaC_2 \rightarrow 2\,(CaS) + CO\,(gas)$$
or
$$Mg + [S] \rightarrow (MgS)$$
(Square brackets, i.e. [S], show that material is dissolved in the hot metal. Round brackets, i.e. (CaS), show material dissolved in slag.)
– Removal of carbon, silicon, manganese and phosphorous. These elements react with the oxygen blown into the converter. The "affinity" for oxygen decreases in the sequence Si>Mn>C>P>Fe. In this sequence material is refined in the converter process. At the end of the refining process iron can be reoxidised, which is sometimes required to heat up the steel before casting. Si, Mn, P and FeO are removed with the slag phase, the C as CO or CO_2 in the gas phase.

The important considerations for a steel plant are:
- Consistent quality: the control of the converter process incorporates "learning", which adjustments to the process settings are necessary on the basis of expected outcome versus the actual outcome. The more consistent the iron quality, the better the results in the steel plant.
- Hot metal silicon, manganese, titanium and temperature are important energy sources for the converter process and effect the slag formation.
- Hot metal phosphorous has a major influence on steel production process. In the blast furnace 97 to 98 % of the phosphorous leaves the furnace with the hot metal.
- Hot metal sulphur is a problem because sulphur is difficult to remove in the converter process. For high grades of steel a maximum sulphur level of 0.008 % is required, while the blast furnace produces hot metal with a content of 0.030 % and higher. Therefore, an external desulphurisation step is often required.

9.2 Hot metal composition

The final hot metal composition is the result of a complex process of iron–slag interactions as the various elements are divided over the slag and iron phases. The dispersion of an element over the two phases depends on the slag and hot metal composition as well as temperature, as discussed below. As an illustration the typical percentages of elements entering the slag and iron phases are indicated in Table 9.2.

The following points should be noted:
- Silicon, titanium and sulphur are concentrated in the slag.
- Manganese is concentrated in the hot metal.
- Some of the potassium is discharged from the top.
- Nearly all the phosphorous goes to the hot metal.

Element	Input kg/tHM	Output Iron kg/tHM	%	Output Slag kg/tHM	%
Silicon	46	5	11 %	41	89 %
Manganese	6	4.5	75 %	1.5	25 %
Titanium	3	0.7	23 %	2.3	77 %
Sulphur	3	0.3	10 %	2.7	90 %
Phosphorous	0.5	0.48	96 %	0	0 %
Potassium	0.15	0	0 %	0.11	73 %

Table 9.2   Typical distributions of selected elements over iron and slag

9.3 Silicon reduction

Silicon, manganese and phosphorous oxides are reduced via the direct reduction reaction. Out of these three, the silicon reactions are of particular interest. The hot metal silicon is a sensitive indicator of the thermal state of the furnace, and the silicon variation can be used to analyse the consistency of the process. For these reasons the silicon reactions are discussed in more detail.

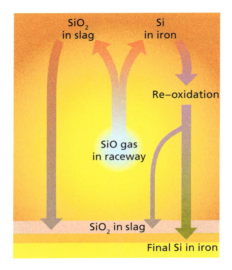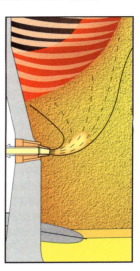

Figure 9.1   Reactions of silicon in the blast furnace

The reduction of silicon takes place via three steps (Figure 9.1):
- Formation of gaseous SiO in the raceway. The first reduction step takes place at the very high flame temperatures of the raceway. The silicon comes from the ash of the coke (and coal). The higher the coke ash, the higher the silicon in hot metal.
- Further reduction by means of direct reduction with the iron. The SiO gas in contact with the iron can be reduced as follows:
  $SiO + [C] \rightarrow [Si] + CO$
  (square brackets indicate solution in iron).
- The more intimate the contact between iron and gas, the higher the hot metal silicon content. The higher the height that the iron drips down, the greater is the contact between the hot gasses and the liquid metal, leading to higher hot metal temperatures. The longer contact allows more SiO gas to react with the carbon in the hot metal, leading to higher hot metal silicon content. Therefore, a high–located melting zone corresponds with high hot metal temperature and high hot metal silicon.
- The hot metal silicon is in equilibrium with the slag. Important aspects are:
- When iron droplets descend and pass through the slag layer, the silicon can be reoxidised if FeO is present in the slag, according to:
  $[Si] + 2 (FeO) + 2 [C] \rightarrow (SiO_2) + 2 [Fe] + 2 CO$

- The more basic the slag (less SiO_2 in slag), the lower the hot metal silicon.
- The hot metal formed in the centre has high silicon, while the hot metal formed at the wall has low hot metal silicon. The cast result is an average value.

Hot metal silicon and manganese are both indicators of the thermal state of the furnace. Manganese shows a quicker response on process changes due to the fact that the equilibrium with the remaining slag in the furnace is faster for manganese due to the smaller fraction of manganese in the slag.

9.4 Hot metal sulphur

The hot metal sulphur is governed by the input of sulphur, the slag composition and the thermal state of the furnace.

The most important parameters are:
- Sulphur input: the sulphur input is typically 2.5 to 3.5 kg/tHM. The main sources being coke and the auxiliary reductant such as coal or oil.
- The division of sulphur between iron and slag, indicated by the (S)/[S] ratio. This ratio is very sensitive to the slag basicity and the thermal level of the furnace (hot metal temperature or hot metal silicon).
- The slag volume: the lower the slag volume per tonne hot metal, the higher the hot metal sulphur at the same (S)/[S].

Most companies have their own correlations between (S)/[S] and the slag basicity and thermal level. The correlations are derived on the basis of historical data for a blast furnace. As a basic guide: to reduce hot metal sulphur by 5 %:
- reduce input by 5 %.
- Increase basicitiy by 0,02 (basicity defined as $CaO+MgO/SiO_2$) or
- Increase hot metal silicon by 0.06 %.

9.5 Slag

9.5.1 Slag composition and basicity

Slag is formed from the gangue material of the burden and the ash of the coke and auxiliary reductants. During the process primary slag develops to a final slag. Composition ranges are presented in Table 9.4. Four major components make up about 96 % of the slag, these being SiO_2, MgO, CaO and Al_2O_3. The balance is made up of components such as manganese (MnO), sulphur (S), titanium (TiO_2), potassium (K_2O), sodium (Na_2O) and phosphorous (P). These components have a tendency to lower the liquidus temperature of the slag. The definitions of basicity are given in Table 9.3.

B2	CaO/SiO_2	
B3	$CaO+MgO/SiO_2$	
B4	$(CaO+MgO)/(SiO_2+Al_2O_3)$	

Table 9.3 Definitions of basicity (weight percentage)

	Typical	Range
CaO	40 %	34–42 %
MgO	10 %	6–12 %
SiO_2	36 %	28–38 %
Al_2O_3	10 %	8–20 %
Total	96 %	96 %

Table 9.4 Typical slag compositions

9.5.2 Slag properties

Slag has much higher melting temperatures than iron. In practice it is more correct to think in temperature ranges than in melting points, as composite slags have a melting trajectory rather than a melting point. At the solidus temperature the ore burden starts melting. The liquidus temperature is the temperature at which the slag is completely molten. At temperatures below the liquidus temperature solid crystals are present. These solid crystals increase the viscosity of the slag. In our experience the behaviour of slag can be well understood on the basis of its liquidus temperature. Liquidus temperatures are presented in ternary diagrams as shown in Figure 9.2.

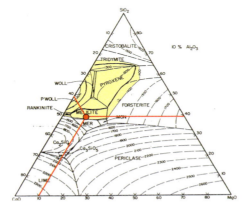

Figure 9.2 Phase diagram of liquidus temperatures of blast furnace slag system for 10 % Al_2O_3. The slag composition 40 % CaO, 10 % MgO and 36 % SiO_2 is also indicated. To this end the components have to be recalculated from 96 to 100 % of the slag. The area where the liquidus termperature of the slag is lower than 1400 °C is indicated in yellow. (After slag atlas, 1981.)

These diagrams have been developed for pure components and in practice the liquidus temperatures are somewhat lower. Since in the ternary diagrams only three components can be indicated, one of the major slag components is taken as fixed. i.e. Al_2O_3 content is 10 %. Diagrams at different Al_2O_3 percentages are presented in Figure 9.3. The typical slag composition for a blast furnace slag is also indicated (Table 9.4). Note that the liquidus temperature is about 1400 °C and that the liquidus temperature increases when CaO increases (i.e. when the basicity increases).

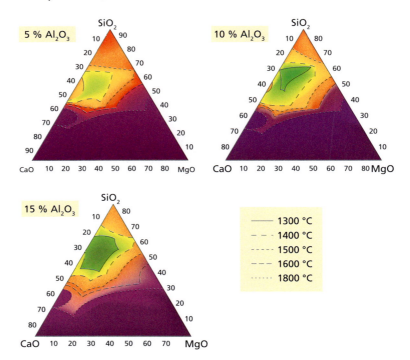

Figure 9.3   Phase diagrams of slag liquidus temperatures at various Al_2O_3 levels. (After slag atlas, 1981.)

In Figure 9.4, the composition of the slag resulting from a burden of self fluxed sinter and pellets is indicated. The liquidus temperatures of the "pure" components give high liquidus temperatures for the slag, well above 1500 °C. How is it possible that the material melts in the cohesive zone?

The secret behind the melting of sinter and pellets is, that the ore burden contains a lot of FeO, which lowers the melting temperature or, as mentioned earlier, lowers the liquidus temperature and solidus temperature. This is indicated in Figure 9.5. Here, the diagram of CaO, SiO_2 and FeO is presented. At a basicity (CaO/SiO_2) of 0.9 the liquidus temperature of slag decreases, when FeO is present. At 0 % FeO, the liquidus temperature is 1540 °C, at 20 % FeO it's 1370 °C and at 40 % FeO it's 1220 °C. In the presence of Al_2O_3, the effect is even more pronounced and FeO can lower the slag liquidus temperature to about 1120 °C (data not shown). The primary slag, i.e. the slag formed during melting process and prior to solution of the coke ash components into the slag, is made liquid due to dissolved FeO.

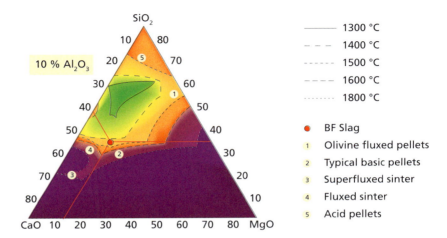

Figure 9.4   The slag composition of typical pellets and sinter qualities

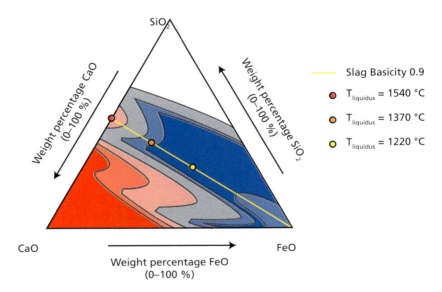

Figure 9.5   Influence of FeO on slag liquidus temperature

The final slag is made liquid through the solution of SiO_2 as indicated in Figure 9.6. The SiO_2 dissolves in the slag during it descent to the hearth.

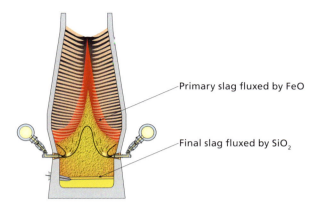

Figure 9.6   Slag formation

9.6 Hot metal and slag interactions: special situations

During special blast furnace situations like a blow–in or a very hot furnace the hot metal silicon can rise to very high values. Since the silicon in the hot metal is taken from the SiO_2 in the slag, the consequence is that the basicity increases. This leads to high slag liquidus temperature (Figure 9.7).

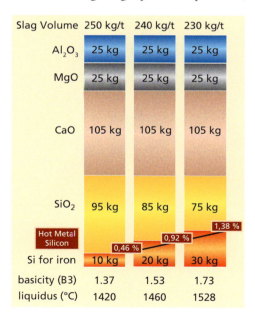

Figure 9.7   Slag properties if hot metal silicon increases, a typical example

In a situation with very high basicity the final slag is no longer liquid in the furnace and cannot be cast. It will remain in the furnace where it can form a ring of slag, particularly in the bosh region. Burden descent and casting will be disrupted. Therefore, for special situations where hot metal silicon is expected too be high, the slag should be designed to handle the high hot metal silicon. To this end, extra SiO_2 has to be brought into the furnace and the recommended method is the use of siliceous lump ore.

Some companies use quartzite, which is suitable to correct the basicity in normal operation however, it is not suitable for chilled situations, since the liquidus temperature of quartzite itself is very high (1700 °C). The effect of the use of a siliceous ore can also be shown in the ternary diagram in Figure 9.8: by working at a lower basicity, the liquidus temperature decreases along the indicated line.

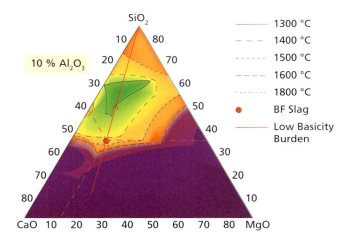

Figure 9.8   Effect of low basicity burden on slag liquidus termperatures

X Casthouse Operation

10.1 Objectives

The casthouse operation is an extremely important area for the blast furnace. The main objectives of good casthouse operation may be summarised as follows;
– To remove liquid iron and slag from the furnace at a rate that does not allow the process to be affected by increasing liquid levels in the hearth
– To separate and sample the iron and slag that is cast from the furnace
– To direct the iron to the ladle and the slag to the slag pot, pit or granulator

The extraction of liquids from the hearth is crucial for maintaining stable process parameters, and the damaging effects of not casting the furnace will very quickly become apparent. In this chapter the link between casting and the Blast Furnace process will be explained, and the factors that determine the ability to cast the furnace are discussed.

10.2 Liquid iron and slag in the hearth

The blast furnace process results in liquid iron and slag being produced. These two liquids drip down into the coke–filled hearth of the blast furnace where they wait to be tapped, or cast, from the furnace. The densities of the two liquids are quite different; with iron (7.2 t/m³) being three times that of slag (2.4 t/m³). This difference leads to very good separation between the iron and the slag once it is outside the furnace, given the correct trough dimensions, but also means that separation will occur inside the hearth before the liquids are tapped, see Figure 10.1.

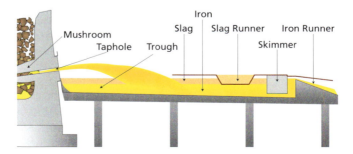

Figure 10.1   Slag and iron separation in the iron runner, or trough

The trough will still hold liquids from the preceding cast, so when the iron from the next cast starts flowing, it will then increase the level in the runner so that the iron already under the skimmer will also increase in height and start flowing again over the iron dame. This iron will then flow to the tilting runner and into a torpedo ladle. Once the ladle is full, the tilting runner will be repositioned into a torpedo ladle which is parked alongside the full one, for that also to be filled. The full ladle will be changed in the meantime for an empty one, so that the cast is not interrupted.

The slag is sitting on top of the iron, so it does not flow under the skimmer so long as the separation remains good. Once it has reached a certain level in the trough it will flow over the slag dam and to either slag granulator or to a slag pit or ladle. It is very important that iron is not allowed to go down the slag dam as this can result in explosions in the granulator, or difficulties in emptying the slag pit. For yield reasons it is also not desirable to have slag going into the torpedo ladle.

The hearth itself is a refractory vessel contained by the steel blast furnace shell, as shown in Figure 10.2. Cooling of the steel shell is essential to avoid overheating of the refractory and shell to the point of failure. The taphole or tapholes are positioned such that a pool, or sump, of liquids remains in the bottom of the hearth to protect the pad, even after casting. The lower part, known as the salamander, is only tapped at the end of a campaign, to allow for access to the pad for demolition and replacement.

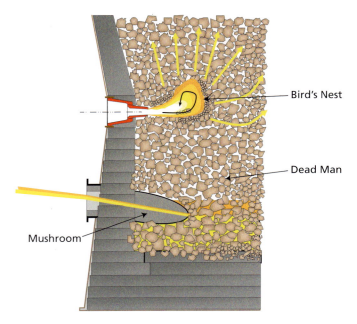

Figure 10.2 The Blast Furnace Hearth

10.3 Removal of liquids through the taphole

The regular removal of liquids from the hearth is done through the taphole, or tapholes. The number of tapholes can range from one to five, depending on the size and output of the furnace. The majority of modern high productivity blast furnaces have been between 2 and 4 tapholes. In normal operation of a furnace with two or more tapholes, the tapholes will be used alternately, with one cast being on one taphole, and the next cast being on the other. This also applies to furnace with up to five tapholes. The reason for the extra tapholes is to ensure that there are always two tapholes in operation, even through times of casthouse repair, or emergency breakdown.

The tapholes are openings in the Blast Furnace shell with special refractory constructions built into the hearth sidewall. The tapholes are opened by either drilling through the refractory or by placing a bar in the refractory that is later removed. The holes are closed by forcing a plug of malleable refractory clay into the hole, which quickly hardens to securely seal the hole. In normal operation this taphole clay will extend into the hearth, forming a taphole mushroom that will protect the original refractory construction (see Figure 10.3).

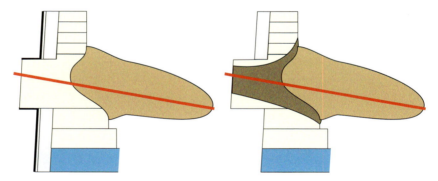

Figure 10.3   Over the taphole campaign, the original lining will gradually be worn away and replaced by taphole clay

The tapholes are perhaps the most vulnerable areas of the blast furnace due to the constant wear and tear and reliance on consumable materials, equipment, and manual intervention. If any of these factors are performing less than optimally, then a deterioration in the taphole performance is the likely result.

The common taphole degradation causes are listed below;
– Improper (e.g. not central) drill positioning when opening the taphole
– Manual oxygen lancing to open the taphole
– Clay leakage out of the taphole on closing the hole
– Water leakage from inside the furnace
– Gas leakage through refractory surrounding the taphole itself
– Slag and iron attack – both chemical and physical

The liquid iron and slag flow from the taphole are determined partially by the flow to the taphole on the inside of the hearth, but also by the characteristics of the taphole itself, such as:
- The length of the taphole, which is affected by the plugging practise and the clay quality
- The diameter of the taphole, both the diameter at which it was opened, but more the wear of the taphole over time
- The roughness of the surface of the taphole
- The pressure inside the furnace, consisting of the furnace blast pressure and the liquid hydrostatic pressure

As the taphole will wear through the cast, especially when slag starts to flow, the rates of iron and slag flow are not constant through the cast. Even with good casting regimes there will be a some variation in the hearth liquid level, with the desired situation being as little variation as possible. The taphole clay quality determines the resistance to slag attack, and therefore the choice of clay quality is very important. This is often determined by availability of local supply, and so is not discussed in detail here.

The length of the taphole is determined by the amount of clay injected, and so more clay is always injected than is needed to just close the taphole. The excess clay is pushed beyond the end of the taphole and forms a 'mushroom' at the opening of the taphole in the hearth itself. This mushroom protects the taphole block itself from wear. The larger the furnace, the bigger the mushroom inside the hearth, and so the longer the taphole. An 11 metre furnace can expect to have a taphole length of 2.5 m minimum, and at 14 m hearth diameter this increases to 3 m.

10.4 Typical casting regimes

A blast furnace will be cast between 8 and 14 times per day. These casts may last between 90 and 180 minutes, with the end of the cast indicated by a spraying of the liquids caused by gas from the raceway escaping out of the taphole. In this time the furnace processes a considerable part of its working volume. As shown in chapter 2, the residence time of the burden is approximately 6 hours. Therefore a 2 hour cast represents a third of the content of the blast furnace being transformed from burden material to molten iron and slag.

Figure 10.4 shows an example of regular tapping sequence using two tapholes. Most two, three and four taphole furnaces will operate in this way, with the extra tapholes being either a spare or out for maintenance.

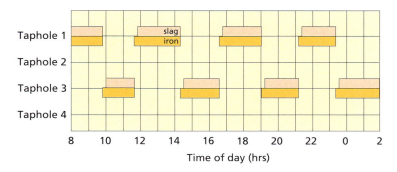

Figure 10.4 Typical casting regimes with a two taphole furnace, showing iron run times with slag above them

When the tapholes are closed, or one is open but the stream of liquid exiting has a low flow rate, then the liquid level in the hearth will increase. That is to say, the production rate is higher than the tapping rate. If this continues for long enough, then the increased liquid level in the hearth can affect the blast furnace process in the following ways:

1. The upward force on the submerged coke deadman is increased by the increased liquid level. This increase in the upward force will slow down the burden descent.
2. If the slag level is so high that it reaches the tuyeres then the gas flow will be severely affected, with increased gas flow up the wall. This can result in poor reduction of the burden and therefore a chilling furnace.
3. The slag can be blown high up in the active coke zone, impeding normal gas distribution
4. If the hot metal level is so high that it reached the tuyeres, then it is possible a cut tuyere will be the result, causing water leakage into the furnace. In the worst case scenario the tuyere will burn severely or a blow–pipe will fail. This will then lead to a blow–out of coke and a very critical emergency stop.

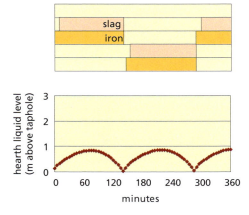

Figure 10.5 Casting and Hearth Liquid Level

In order to avoid any of these effects, the hearth liquid level should be kept under control and preferably at a low level, as per the example given in Figure 10.5. In a modern, high productivity blast furnace, measurement of hot metal and slag quantities should be registered real time, so that the casthouse crew can take timely actions.

10.5 Taphole drill and clay gun

These two pieces of equipment are two of the most critical items on the blast furnace. The maintenance of these items must be of a very high standard as the availability of them on an active taphole can not be any less than 100%. Cleaning of the gun nozzle after every plug is essential for ensuring that the clay can be pushed at the next cast, which in turn will prevent the gun nozzle being burned.

It is important to keep the taphole face clean and to clean down the sides of the trough regularly so that there the mud gun can swing into place without obstruction and the nozzle gets a good seal on the taphole face.

The clay quality and method of plugging the hole with the clay are very important for both the length of the taphole and the flow rates of iron and slag. Plugging has to be done at the same position as the drill has opened the hole to avoid clay spillage.

The speed of the piston and the pressure used to force the clay into the hole has a strong influence on the ability of the clay to plug the taphole effectively. If the clay can only partially fill the hole then the next time the cast is opened the drill will have more difficulty in opening the hole as it is also trying to cut through iron particles. This is one of the reasons why the production rate of the furnace can be limited by the taphole equipment, and so serious consideration should always be given to upgrading the taphole gun and drill whenever significantly higher production rates are targeted.

To preserve gas tightness of the taphole the post–pressing technique can be applied. This technique involves pressurizing the clay with the clay gun after it has filled the hole, to try and close any small cracks or fissures in the taphole.

Ensuring that the taphole drill is in the centre of the taphole each and every time is also very important as otherwise the gun will not be able to plug the taphole as well as it should, leading to less clay going in the hole and so a shortening of the taphole and also potentially burning the gun. A selection of drill bit diameters can be used, although the aim diameter should be kept relatively constant when aiming for consistent tapping practises. The range of drill diameters is then useful for special situations, when the tapping is irregular, or changes to the production rate requires changes to casthouse practise.

As an alternative to the drill, a soaking bar may be used. This is a bar of solid steel that is hammered through the clay immediately after it has been pushed into the hole, while it is still soft. The clay is then allowed to harden and the bar is pulled out. This results in a very smooth taphole of equal diameter throughout, although the hammering of the bar in and out of the taphole can increase the stresses on the taphole block construction itself and introduce gas leakages.

10.6 Hearth liquid level

The level of liquids in the hearth should always be kept as low as possible. This means that the hearth should never be used as a 'buffer' for the containment of produced liquids. The reason for this is that the liquid level, above a certain level, has a direct impact on the process. As shown earlier in Section 7.2, the liquids in the hearth act as an upward force in the blast furnace, along with the blast pressure. Should this force be allowed to increase, it will impact on both the blast pressure and the descending burden. It is shown schematically in Figure 10.6 what happens in the furnace when the liquid level increases too far.

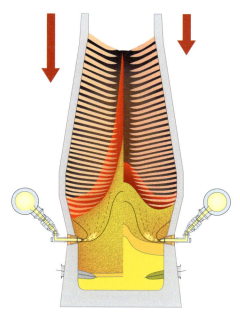

Figure 10.6   Consequences of increased liquid level
(Arrows indicate burden descent rate)

As shown, the high liquid level causes the blast to be deflected more towards the wall, rather than through the centre of the furnace. This is because the coke in front of the tuyeres has been infiltrated with slag, and so is much less able to accept the flow of the gasses produced at the raceway.

In this instance the bosh is subject to much higher heat loads than normal, and the root of the cohesive zone will increase. However, at the same time the centre of the furnace the cohesive zone will drop due to the reduction in gas passing through the centre.

The blast pressure will also be higher as the resistance in front of the tuyeres is higher, and the burden descent will slow considerably. The furnace may even begin to hang, with the danger of slag filling the tuyeres should the furnace then slip, where material will quickly drop into the full bath of liquids.

The wall temperatures all the way up the stack will also increase, as the gas continues to preferentially travel against the furnace wall. This then subjects the cooling elements to a higher heat load than they will usually encounter. This increase in heat losses, coupled with the loss in furnace efficiency can lead to cooling of the furnace. In this scenario the furnace should be cast without delay, and actions taken to restore the process stability.

Figure 10.7 shows the effect on stockline level in the case where high residual liquid levels is affecting the burden descent. The burden descent slows when the taphole is closed, and then speeds up significantly towards the end of cast, to the extent that the charging system is unable to keep up and a lowered stockline is the result.

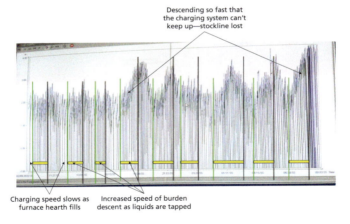

Figure 10.7   *High residual liquid levels and burden descent*

10.7 Delayed casting

In most plants the casting regime will have been calculated and observed to arrive at an optimum length of time in between casts. This is referred to as the gap time, defined by the time between stopping liquid flow by closing one taphole and starting liquid flow by opening another, or in the case of single taphole furnaces, reopening the same taphole. This will be determined by the production rate, number of tapholes, and casting rate. In the majority of cases this casting regime will be adhered to, that is to say, the gap time will be met. However, where there are problems in meeting this schedule, remedial actions may be required.

When casting the furnace it is required to have a good, controlled liquid flow rate from the furnace. Where a taphole is open but is not casting well, the flow should be improved by, for example, re–drilling the hole or re–drilling with a larger drill bit. If the slow flow is allowed to continue then it is quite possible that the furnace will be producing liquids at a higher rate than they are being cast, which will lead to problems inside the furnace.

Whether the casting is delayed, or indeed the casting speed is slower than the production speed, one of the factors that effects the filling rate of the hearth in terms of height is that of the coke bed voidage. The coke bed voidage is an unknown value. Studies have shown that it can vary between 20% and 30% but as yet there is direct method of measuring it. It is also quite likely that the voidage of the coke bed will vary between the centre and the peripheral, and from the bottom to the top, so the assumed overall voidage is not directly applicable to every area in the coke bed. The coke quality will have a strong impact on the voidage, as the breakdown of the coke higher up in the furnace will generated fines, and a wider size distribution of particles that will create a more densely packed coke bed.

By way of an illustration of filling speed, take for example an 8.5 m diameter hearth blast furnace, with a taphole to tuyere distance of 2.6 m producing 3630 tonnes per day with a slag rate of 220 kg/tHM. By calculating the volume of space between the taphole and tuyeres, assuming a coke bed voidage of 20 %, the length of time until the liquid level is at the tuyere can be calculated. In this case it is 62 minutes. If the coke bed voidage is 25 %, then this increases to 77 minutes, and at 30 % voidage it is 93 minutes. We therefore have the situation whereby in one instance the furnace has 90 minutes of full production before the hearth liquids are at tuyere level, and another instance when it has only 60 minutes.

Once the liquid level is at the tuyere, it is already expected that problems with blast pressure will have been experienced, so actions may already have been taken to reduce the blast volume. However if the problems that caused the delayed casting are not resolved when the furnace has already reached this stage, then it will become impossible to take the furnace off wind without slag, and even iron flowing into the blowpipes.

For these reasons it is considered to be good practice to take remedial actions immediately when it is known that the casting will be delayed, regardless of the reason. Estimates may be given for the completion of work, or the restoration of services, but as far as the blast furnace is considered it will continue to produce iron regardless, and if the original estimates are found to be wrong, it will often be too late to take anything than extreme reactions to try to protect the blast furnace. If the iron, and more importantly the slag, is not removed from the furnace in a timely manner, then the process will very quickly suffer, with the extreme case being a frozen hearth.

In the case where the operator is faced with a casting delay, different actions may be taken depending on the current condition of the blast furnace. If it is still casting the previous cast, and it is safe to continue to do so, then the oxygen and then wind rate may be reduced prior to closing the hole, reducing the production rate and so giving a much longer safe gap time.

In this situation the action to reduce production rate should be aimed at safe operation continuing, for example, wind rate should be reduced to the minimum at which injection remains on the furnace. Oxygen should be decreased to the minimum, determined by a simple formula, such as for every 30 kg/tHM injection over a limit of 70 kg/tHM the oxygen enrichment should be increased by 1 %.

Due to the uncertainty in the available voidage for hot metal and slag, it is prudent to make conservative estimates when determining the control actions to be taken.

10.8 No slag casting

As the iron is below the liquid slag, and the taphole elevation will always be at the depth of the iron pool at the start of cast, then iron will be cast before the slag. As the liquid level drops, then a mixture of slag and iron will begin to flow. At the end of the cast the majority of liquid is slag, with iron flowing at the production rate. Sometimes, however, the furnace will cast iron without casting slag, or at least not as much as should be cast.

Although the iron is the focus of the blast furnace, the iron cannot be made without the slag, and due to the nature of it, the slag proves to be the more difficult liquid to cast. Basic slags have a higher melting temperature than acid slags, but the basic slags are more desirable for the desulphurisation properties, so for hot metal quality it is required to use a more basic slag. In time of difficulties, however, one of the first actions to ensure that the furnace will be able to cast well is to reduce the slag basicity. This will give the operator the best chance of being able to get the slag out of the taphole.

If events in the furnace cause a change to either the temperature or the composition of the slag, then it can become much more viscous than the iron, and drainage through the coke bed becomes increasingly difficult. The iron will flow much more easily, and so it can occur that casting will continue with little or no slag being cast. The slag is still being produced, however, and so it is very important to make sure it comes out of the furnace before it interferes with the process.

The problem may be seen to be developing at an earlier stage by monitoring the following parameters:
– Amount of slag cast, measured by the number of slag pots filled or by indirect methods such as the speed at which the slag granulator drum rotates, or temperature pick–up in the granulator outlet water.
– Percent slag time – this is the number of minutes that slag has been cast divided by the number of minutes in the cast, expressed as a percentage. Ideally this number should be fairly constant and representative of the slag volume that the furnace is producing, however it is only accurate when the flow of slag is constant between casts.
– Slag over time – this is the point in time when the slag first flows over the slag dam. Slag will have started exiting the taphole before this point, but not in large enough quantity to give a good indication.
– Slag Gap – this is the number of minutes from when the liquids stopped being cast at the end of the previous cast to the slag over time of the current cast.

When it is clear that the slag is not draining from the furnace as well as it should be, efforts should be made to improve the slag drainage. This may be done by a variety of methods, and it is likely that procedures already exist for it. Using a larger diameter drill bit on the next cast will increase the flow, and may improve the situation. If the taphole is already short, however, and a short cast caused the lack of slag, it may be better to increase the length of the hole so that a longer cast is the result. The problem may only be at one taphole, so changing to the other taphole will already improve the situation inside the furnace. Opening the second taphole should be done after a defined period of no slag casting, as specified in the standard operating procedures for the plant.

If the furnace is on a cooling trend, combined with difficulties tapping slag, increasing the fuel injectant to warm up the fresh iron and slag may temporarily improve the situation, but a coke rate increase will also be required.

Shortening the gap time may also be advisable, especially when it is suspected that liquids remain in the furnace.

10.9 One–side casting

Furnaces with only one taphole are of course optimized for tapping single sided, as are some blast furnaces that follow a routine of having one taphole in operation and one as standby. The majority of two and more taphole furnaces, operate on an alternating taphole basis using two tapholes. This will mean tapping through one taphole, closing it, and then either opening the second taphole immediately or waiting the designated gap time before opening the hole.

The single most important effect of single taphole casting compared to alternate casting is that of the gap time. During the gap time the furnace is still producing liquids but not casting them. Ideally the gap time is calculated as the optimum to allow enough liquid accumulation in the hearth to allow a smooth cast for the desired period of time, with good iron and slag removal, but without increasing the hearth liquid enough to affect the blast pressure. However the gap time can also be affected by external factors such as how long it takes to change torpedoes, clay cure time, maintaining and cleaning the runner system, etc. Where this is the case then it is very important to remember that the furnace is still producing liquids at the same rate, unless a change is made to slow down the production, see Figure 10.8.

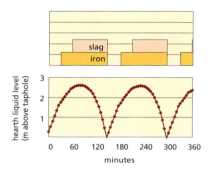

Figure 10.8   Effect of single taphole casting on hearth liquid levels

In single taphole furnaces the minimum gap time is often dictated by the curing time for the clay. If the taphole is opened before the clay has hardened, much of it will easily wash away, which will quickly erode the taphole mushroom and expose the taphole refractory block itself. With alternating casting this is not a problem as the clay has the time that the other taphole is in use to harden. Therefore, single taphole furnace use resin bound clay types.

The gap time has major impact on hearth liquid level and thus on the process results. In Figure 10.9 the effect of the gap time on hearth liquid level is simulated: it is clear from the figure, that in this calculation the highest hearth liquid level rises from 2.5 m above taphole to 3.8 m above taphole when gap time is increased from 30 to 60 minutes.

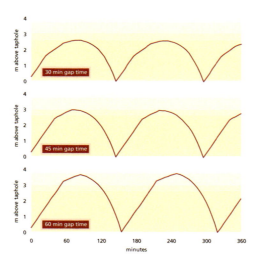

Figure 10.9 Effect of gap time on hearth liquid level, single taphole operation.

If a furnace must switch from alternate to single sided casting the area to look at firstly is the difference in gap time between the two practices. If alternate casting requires a gap time shorter than the time it takes for the clay to harden, then single casting will require a change in practice. If different clay is available, then this may be applied, but caution should be used during the transition as the clay already in the hole may not combine well with the new clay.

If there is a significant difference in the gap time then to minimize the fluctuation in hearth liquid levels, it may be advisable to reduce the production rate. Experience has shown that an 11m hearth diameter blast furnace can produce 5500 to 6000 t/d with one taphole, and a 14m heath diameter furnace can produce around 8000 t/d. This is often a significant reduction compared with what the furnace is usually producing.

10.10 Not dry casts

A cast that has ended before all the liquids have been drained from the hearth is described as a not dry cast. This is reported whenever the taphole has to be stopped during a cast, such as when the torpedoes are full, or there has been a problem in the casthouse that required the flow of liquids to be stopped. Other causes can be a very short taphole or a crack in the taphole mushroom. It is good practice to record the suspected reason for a not–dry cast so that improvement plans for the worst offenders can be made.

A not dry cast may also be reported when the taphole is showing signs of end of cast, when it can be reasonably suspected that the furnace is not empty. This could be when the slag is not yet over, or it has only been casting for a very short time, or not enough liquid volume has come out of the furnace.

A third example of a not dry cast is more difficult to determine, and can easily be missed as the signs are less obvious and may only be picked up in the control room, rather than on the casthouse itself. In the case of a series of casts where the casting has appeared to be normal, it is still possible that some slag has been retained in the furnace after each cast. This will not be noticed after one or two, or depending on the amount, perhaps even more casts, but after successive casts where a small amount of slag has been retained in the furnace, it will build up to a large amount. At the point the blast pressure can begin to be affected.

This will be more noticeable when the furnace is closed as the blast pressure may increase, and continue to increase until the taphole is opened again. It may not decrease again until the slag begins to tap at a reasonable rate, and so lowering the level in the furnace. As the signs with blast pressure are not always a precise match with the casting times it can sometimes be dismissed as the cause. On these occasions it is useful to look to the slag time percentage, as well as the slag run durations themselves.

Depending on the cause of the not dry cast, slightly different reactions may be appropriate. Where the not dry cast is known and the taphole is closed for operational reasons, the second taphole should be opened immediately. Where this is not possible the oxygen and then wind rate should be reduced and the original taphole is re–opened as soon as possible. Where this is not possible, the decision to close the taphole should be delayed as much as possible, with wind rate being reduced as far as liquid levels will allow. At this point it is a balance between how much damage is being caused outside the furnace due to, for example, molten metal spill, compared to the danger of flooding tuyeres with slag and iron.

In the case where the taphole has shown signs of the hearth being empty, but it is thought that it is not from the casting times and amount of slag cast, then there are a few different actions that may be considered. If there is a second taphole available then it may be opened prior to the first being closed. Once this is safely open the first one may then be closed, known as overlap casting. Alternatively, the normal gap time between casts may be reduced to zero, so the second taphole is opened immediately after the first is closed. It is important to ensure that both tapholes do not finish casting at the same time as that will introduce a necessary gap time, so once slag appears at one of the tapholes, it should be closed to allow the other to cast normally. This technique of when to open and when to close a second taphole should be included in the standard operating procedure for casting to ensure that the best sequence, proven in practice, is followed by all operators.

In either case, a larger drill bit may be used to open the original taphole again, when it is due to cast. This may help in removing the liquids from this side, assuming that a short taphole length is not the cause of the problem.

Where only one taphole is available, the taphole may be closed for either a much reduced gap time, for example 10 minutes rather than 30 minutes, with a shorter clay stop. It is also possible to stop the taphole without clay for a minute or so, but it should first be checked whether the gun is sufficiently protected to do this. This practice should not be repeated on the same taphole as it will allow the taphole mushroom to erode too quickly, causing further problems.

These same actions may also be taken if the blast pressure is being affected by a possible build up of slag in the furnace. At the same time, however, other causes of increasing blast pressure should also be investigated.

10.11 Defining a dry hearth

Witnessing a blow at the taphole is often considered to be the definitive critera for whether the furnace is dry or not. Although a good indicator, and should never be taken for granted, a blow at the taphole only indicates that the liquids in the vicinity of the taphole are drained, and does not say anything about liquids in other areas of the hearth. Where the drainage to the taphole is poor from area far from the taphole, then it is possible for liquid levels in the area of the taphole to drop sufficiently low for a blow at the taphole to appear while there are still a lot of liquids left in the furnace. In this scenario the taphole should still be plugged, but the cast is to be considered to be a not dry cast. Unfortunately these are not always possible to determine from the casthouse. The indicators of a dry hearth can be summarised as the following;

1. Casting until a blow at the taphole is witnessed
2. Enough slag and iron has been removed from the furnace to correspond with the known production rate
3. The process parameters show no sign of the hearth holding liquids – blast pressure normal, charging rate normal
4. The furnace can be shut down at any time, without concern that slag or iron will flow into the tuyeres.

It is the last of these criteria that is often the defining one, where the decision to take the furnace off for a short stop is delayed until after the next cast. This in itself indicates that the operator is not confident that the hearth has been drained sufficiently to avoid any residual liquid threatening to enter the tuyere when the blast pressure is reduced. An operator who can confidently take the furnace off blast at the end of the current cast is one who has confidence that the furnace is draining well during the cast.

10.12 Oxygen lancing

On occasion it is unavoidable to open the taphole using oxygen lancing. This practice should be considered a last resort as it is extremely damaging to the taphole refractory. Where the use of oxygen lances is increasing, the situation should be investigated very closely to identify the root cause.

Where the use of oxygen lances is unavoidable, they should only ever be used by experienced casthouse workers, following the pre–drilled hole to ensure that the lance is burning in a straight line down the centre of the taphole. If more than one lance is required the interval between the two should be as short as possible, with the practice continues until the taphole is opened. Where this is causing a long delay to the cast, alternative or additional actions such as opening a second taphole or reducing wind rate should be considered at an early stage.

Repeated use of oxygen lances to open the taphole is likely to cause irreparable damage to the taphole area, and may even pre–empt a taphole break–out or necessitate an extensive taphole repair to avoid such a break–out. There is a very large risk associated with using oxygen lances as it is very difficult to ensure that the lance is burning in a straight line. Damage to the taphole block or to taphole staves are the biggest concern.

10.13 Cast data recording

For good analysis of taphole condition and casting performance it is important to keep very good cast records. Some of the data that should be recorded on a cast basis is as follows:
– Cast Number
– Time start drilling
– Number drills or oxygen lances used to open hole
– Time liquid start flowing
– Drill diameter used to open hole
– Taphole length
– Time slag over
– Time end cast
– Amount of clay used to close taphole
– Clay type used
– Blow at the taphole

The cast end times, drill start times, iron run and slag over times can be plotted very easily to allow quick and easy interpretation of the casting. This method is often much more illustrative and quicker to interpret than the lists of times that are often meticulously recorded. Having the times plotted on a black chart which is being constantly updated, allows problems to be identified very quickly and so solutions applied at an earlier stage than may otherwise have been the case.

XI *Special Situations*

11.1 Fines in ore burden

11.1.1 Segregation of fines and coarse material

The permeability of the ore layer is determined by the amount of fines (< 5 mm) in that layer. Unfortunately, when bulk material is handled, fines are formed. Therefore, normally coke and ore burden are screened before being charged into the furnace.

Moreover, fines tend to segregate. When material is put into stock, the fine material remains on the point of impact and the coarser material roll outwards, known as segregation. This effect is known wherever granular material is handled. So, when reclaiming material from stock, it is important to avoid high amounts of fines being reclaimed and sent to the furnace without screening.

Similar segregation can take place while charging the furnace, and can impact the furnace process. Fines in general are undesirable due to the blocking of the spaces between the larger particles, however due to the flow characteristics of fines, they can also deposit preferentially in certain areas. The impact of this is particularly noticeable with bell–charged furnaces, where the fine particles will drop directly down onto the stockline, and the large particles will flow a little more outward and deposit at the wall (see Figure 11.1). If material hits the wall before it reaches the burden level, the fines will accumulated close to the wall and the coarser material will flow more inwards.

This segregation effect also takes place when filling a bunker, be it in the stockhouse or on the bell–less top, segregation will always take place. When material is required from a bunker, it starts to deliver the material that has been charged in the centre: those being the fine materials, while later the coarser materials from the sides begin to flow.

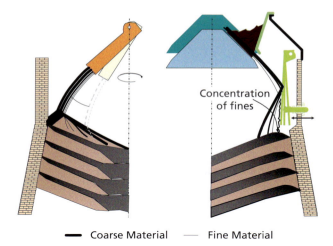

Figure 11.1   Segregation of fines during charging, with a bell and bell–less top charging system

A concentration of fines close to the wall can have a negative effect on the reduction and melting of the ore as it forms a blockage for the passage of hot reducing gasses through the ore layers, as shown in Figure 11.2.

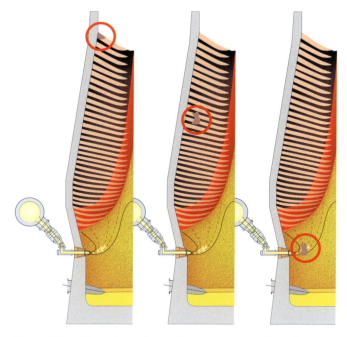

Figure 11.2   Fines charged at wall migrating through the furnace and appearing as 'scabs' in front of tuyeres

Note that there is a difference between the path travelled by the coarse materials and fines. When the burden descends though the furnace, the fines fill the holes as soon as they are formed, while coarse materials follow the wall. Fines travel more vertically and faster towards the cohesive zone! (See Figure 11.2) With a bell top arrangement it is possible to deflect the fines by using the furnace movable armour as a deflector, and with a bell–less top by charging from the outer position to the inner.

An additional source of fines that can be avoided through slight modification in stockhouse practices is that of bin management. The drop that the raw materials experience can vary significantly, depending on the height of the bin. By maintaining a standard bin fill level, such as 65 to 75 %, the quantity of fines generated remains at a constant level. If there are screens after the bins then this will increase the yield and if there are not, it will decrease the fines loading to the furnace.

11.2 Moisture input

The moisture charged into the furnace with the coke and ore burden must be removed before the process can start. This takes place in the upper part of the furnace. The centre dries very quickly, but in the wall area it can take much longer, as shown in the figure, about 40 minutes.

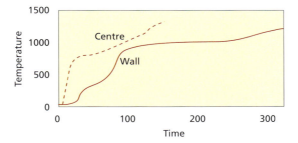

Figure 11.3   Temperature in furnace

If the moisture input increases, then it will take longer for the material to dry and the isotherm where the reduction process will start will descend downwards. As a consequence the reduction process will be less efficient and more oxygen will be removed by direct reduction, consuming energy and so cooling the furnace.

Most companies are equipped with moisture gauges for coke, so that variation of the moisture input in coke is compensated for with an additional weight of coke. Note that this is only a minimum correction to maintain the current thermal state. If the furnace is already in a critical state the compensation with coke moisture gauges will not be sufficient to compensate for the decreased efficiency of the reduction process.

Where moisture is added in place of coke the furnace cools and so the normal thermal control procedures will be activated, usually calling for additional fuel. If the moisture level then reduces again, the furnace will warm up, triggering another set of actions. If this is allowed to continue, the furnace will enter a thermal cycle that will in turn consume more fuel than required, and be at risk of chilling.

This effect is just as important with pellet moisture, especially where pellets have been shipped or stored under damp conditions. They can contain up to 6 % water. When a batch of these pellets are charged to the furnace the top temperature will decrease with the additional moisture, but the furnace will start to warm up due to the lower amount of iron that is being charged to the furnace. Coke rate changes will normally be made to correct for this warm up, however once this batch of wet pellets have been consumed it is very important to realize that the furnace will then cool down due to the additional iron that is being charged. If this is not anticipated then the furnace can cool down very quickly, so it is better to anticipate this change by increasing coke rate when it is known that the wet pellets have been consumed and dry pellets are soon to arrive.

Ideally, coke and pellet moisture gauges can be installed to monitor and correct for any changes on–line. These moisture gauges take regular readings of the as–charged moisture levels for coke and pellets and will make corrections for the weight, so that the required quantity of the material is charged.

The recommended approach is that the top temperature is not allowed to fall for a prolonged period (8–16 hrs) below dewpoint temperature. Some companies are able to run the top gas temperature at low average levels, well below 100 °C. In these situations it is recommend even to monitor the temperatures in the wall area (3–5 m below the burden level) to monitor whether or not the burden is dry 'on time'.

11.3 Recirculating elements

Potassium, sodium and zinc tend to recirculate within the blast furnace. They form gaseous compounds, which condense on colder parts of the burden. These elements can have a negative impact on the refractory condition. Alkalis will affect the coke reactivity (Chapter 3) and in doing so will increase direct reduction reactions.

In furnaces operated with a central gas flow, the top gas temperatures in the centre increase to such a level that part of the alkalis and all the zinc leaves the furnace with the top gas. If top gas temperatures are low, the alkalis and zinc may accumulate in the furnace. The zinc normally condenses on the refractory. Alkali build–up is manifest by observing the potassium content in the slag, especially when the slag is acid and/or the furnace is cold. Alkali leaves the furnace easier with a low basicity ($B_2 < 0.9$) slag and at low HM temperature. One rule of thumb is, that as long as K_2O in a lean or cold cast is < 1%, no significant accumulation takes place. It is also observed how fast the potassium in the slag returns to a normal level, when slag is lean, such as when preparing for a stop.

11.4 Charging rate variability

Most operators observe the charging rate in a furnace as defined by the amount of charges put in the furnace per hour. If the charging rate increases, while tuyere conditions are unaltered, the furnace will fall short of heat. Simply put, with the same amount of heat and gas produced at the tuyeres more hot metal is made, so the furnace will chill. The reasons for this happening can be various; a fuel shortage as a consequence of too low coke input (correction of coke moisture gauges); too much input of ferrous material (e.g. when changing from 'wet' pellets to dry pellets); or by changing process conditions.

Here we refer to increased direct reduction reactions. In some situations the gas reduction of the burden does not progress sufficiently. This can be caused by
– Too much water input, lowering the isotherms within the furnace and shortening the process height of the furnace, especially at the wall.
– Irregular burden descent, causing mixed layers.
– High residual level which affects the normal gas flow through the burden.
– Charging delays causing that the newly charged material to see shorter process height and altering burden distribution.

The resultant material with insufficient pre–reduction will in any case continue to descend to the high temperature region above the tuyeres. When this material starts melting, all oxygen will participate in direct reduction. This consumes coke and since coke consumption drives the production rate, the production rate will increase further. This is a self propagating effect, and will chill the furnace within hours.

Experienced operators equipped with the right tools can observe the increased direct reduction long before the casthouse gives warning of low hot metal temperature. The method to correct the incident is by slowing down the production rate, with extra fuel injection and/or lower blast volume, and by maximizing heat input into the furnace (maximum hot blast temperature and no blast moisture).

11.5 Stops and start–ups

When a blast furnace in full operation is stopped, some of the processes continue. While the blast is stopped, the direct reduction reactions within the furnace continue as well as heat losses to the wall. The consequence is that the temperature of the material in the melting zone is reduced to around

1000 °C, which is the start of the carbon solution loss reaction. The decreasing temperature re–solidifies the melting materials. Therefore, after a stop it takes some time for the burden to start descending. The burden descent restarts as soon as the "old" melting zone is molten (Figure 11.4).

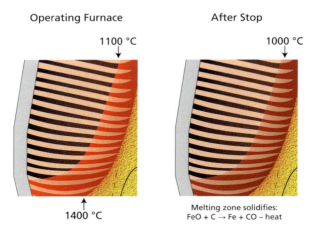

Figure 11.4   Solidified melting zone as consequence of a stop

The heat shortage for a stop of a furnace operating with PCI is even worse: during the stop procedure the coal injection is switched off from the furnace and during the start–up it takes time to restart the PCI. An additional reductant shortage results.

In addition, after a stop the hot metal silicon sometimes rises to very high values, especially if during the stop/start procedure the furnace is operated at a low blast volume. As shown in Figure 9.6, the basicity of the slag will be affected by the high hot metal silicon and might even solidify within the furnace. This results in disturbed burden descent. Heating up the slag is the only solution, which can be achieved by charging extra coke into the furnace 6–8 hours prior to the stop.

So, in order to compensate for the heat losses during a stop and the risk for high hot metal silicon, the following measures have to be applied:
– Extra reductant into the furnace. Coke as well as auxiliary reductants are possible. Additional reductant is needed for a period that the furnace is not operated on PCI.
– Design slag composition for low basicity at high hot metal silicon. Use of a siliceous lump ore is recommended. Even if a stop is unplanned, taking these measures after the stop is worthwhile.

For a blow–in after a stop major pitfalls are:
– Too fast blow–in. The solidified melting zone will take time to melt during the start–up. If allowed time is insufficient, the pressure difference over the burden

can increase too much, leading to gas escaping along the wall (high heat losses) and poor burden descent.
- Too fast restart of the PCI. Since the melting zone is solidified, there is a risk that solid agglomerates will block the hot blast through the tuyere. If this happens, the coal will still be blown into the blowpipe where it can cause blowpipe failure. It is recommended to restart coal injection only when the burden starts descending.
- Too high slag basicity.

11.6 Blow–down

Blowing down a blast furnace requires operating the furnace without simultaneous charging of the furnace. Therefore, all material charged into the furnace is exposed to the same temperatures and reduction processes as if the furnace was fully charged.

However, since the temperature of the shaft gas is not transferred to the cold charge, the off–gas temperatures increases and the gas composition changes. Since the equipment has not been designed to withstand the high top gas temperatures, the top gas temperatures are kept under control by spraying water. The water sprayed above the burden should be prevented from reaching the burden surface, either directly via descent on top of the burden or indirectly via the wall. Special atomising nozzles are required and the success of the blow–down heavily depends on proper spraying. The progress of the blow–down process can be measured from the burden level as well as from the analysis of the top gas composition. Since less and less oxygen is removed from the ore, the CO_2 percentage decreases and CO percentage increases (Figure 11.5).

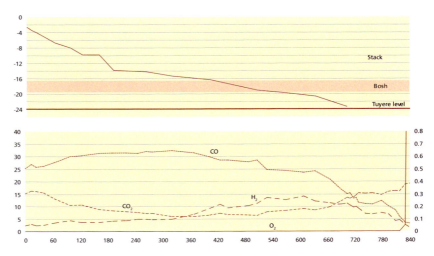

Figure 11.5   Typical progress of a blow–down

Moreover, generally H_2 increases as a consequence of the (unavoidable) contact of spraying water with the hot coke. At the end of the blow–down, when the level of the coke is coming close to the tuyeres, the CO_2 formed at the tuyeres has insufficient opportunity to be transformed to CO and the CO_2 percentage in the top gas increases. As soon as half of the oxygen is in CO_2 (i.e. when the CO_2 percentage equals half the CO percentage), the furnace should be isolated from the gas system.

Normally, a blow–down takes 10 to 12 hours, after a preparatory stop, to reach the tuyere level.

Prior to the blow–down the furnace contains coke in the active coke zone and dead man, and alternating layers of coke and ore in melting zone and stack zone. Since during the blow down the coke of the active coke zone and dead man will be gasified, there is coke excess in the blast furnace. During the latter stages of the blow down reduction reactions have largely stopped, so any auxiliary reductant injection can be stopped during the early stages of the blow down. The moment is indicated by the gas analysis: as soon as the CO_2 percentage starts to decrease to below 10%, then there is little iron oxide left to reduce.

The burden level in the furnace is difficult to measure with standard stock rods. Mechanical stock rods have to be equipped with chain extensions and recalibrated for the purpose. The stock rods should be used only at intervals, since the high temperatures above the burden may cause chain breakage. Radar level indicators can be used if reliable. Indications from the level of the burden can also be obtained from:
– The pressure taps.
– The casthouse operation i.e. the quantity of iron cast.
– Calculation of the amount of coke consumed in front of the tuyeres.

The required condition of the furnace after the blow–down depends on the purpose of the blow–down and consequent repair. Generally the walls have to be clean. Cleaning of the hearth is another important topic. If solid skulls and scabs are expected in the hearth and have to be removed prior to the blow–down, the furnace can be operated for a prolonged period on a high thermal level, relatively low PCI rate and without titanium addition. The effect of these measures is, however, uncertain.

11.7 Blow–in from new

Blowing in a furnace from new can be considered in two phases:
Phase 1 Heating–up the hearth.
Phase 2 Starting the reduction reactions and iron production. The two phases are discussed separately below.

11.7.1 Heating up the hearth

Heat is generated by the reaction of coke carbon to CO. Coke generates 55 kJ per mole carbon, when reacting to CO, which corresponds to 3.9 MJ/ t coke.

The heat requirement in the early stages of the blow–in is for the following:
– Heat coke in the hearth, dead man and active coke zone to 1500°C.
– Heat required for evaporation of moisture from the coke.
– Heat required to compensate for moisture in blast dissociating into hydrogen gas ($H_2O + C \rightarrow CO + H_2$).
– Heat to compensate for loss of heat through the wall.

A detailed analysis of the heat requirement to fill the hearth, dead man and active coke zone with coke of 1500°C indicates the following:
– Moisture in the coke can be neglected.
– The heat required filling the hearth, dead man and active coke zone with hot coke of 1500°C requires an amount of coke gasified to CO of about two–thirds of the estimated volume of the hearth/dead man/active coke zone.
– Additional heat requirement arises from the water dissociation reaction and the heat losses through the wall. For example, if 300 tonne coke is required to fill hearth, dead man and active coke zone with coke, a coke blank is required with a total weight of 600 tonne: 300 tonne to fill hearth, dead man and active coke zone with coke and 300 tonne for the generation of heat to bring the coke to 1500 °C.
– In the early stages of a blow–in, blast temperature should be maximised and blast moisture minimised.
– Heating up the hearth requires some 7 to 8 hours after the blow–in. Heat is generated from coke used at the tuyeres.

11.7.2 Starting the reduction processes

During the early stages of the blow–in while the hearth is heating up, the reduction of the iron oxides has not yet begun due to the temperatures being too low. Therefore, one has to consider the increased amount of direct reduction. The situation may become difficult if the level of direct reduction is too high, (and gas reduction is low). This situation manifests itself from:
– The gas utilisation.
– The direct reduction, as manifest from $CO+CO_2$ exceeds "normal" values.

The gas utilisation is an indication of the amount of gas reduction taking place, while the total CO and CO_2 percentage is an indication for the direct reduction. Especially the CO_2 percentage indicates if gas reduction takes place.

11.7.3 Slag formation

In general, the slag during blow–in has to be designed for high hot metal silicon. However, with the proposed method the hot metal silicon should be under control. If we continue to follow the "two–phase" blow–in approach mentioned here, during the first phase of the blow–in about 350 tonne coke is gasified in 8 hours and the slag formed comes only from the coke ash. Taking 10 % ash and 30 % of the ash as Al_2O_3, we get during the first 8 hours 35 tonne of a high Al_2O_3 slag. This will not cause a problem in the furnace because of the small volume. The coke ash can be diluted, e.g. by using a high siliceous ore in the coke blank. In order to dilute to 20 % Al_2O_3, some 30 tonne of a siliceous ore has to be added to the 350 tonne coke blank.

11.7.4 Hot metal quality during blow–in

As soon as the hearth is heated the hot metal temperature exceeds 1400 °C. As soon as the top temperature exceeds dew–point, all excess moisture has been removed from the furnace and the process has started. There is only limited heat required for heating up and drying of refractories, if compared with the heat requirements of the process itself. So as soon as hot metal temperature reaches 1400 °C and top temperature exceeds 90 °C, the process has to be brought back to normal operation conditions.

However, in this situation the coke rate in the furnace is still very high and the hot metal silicon will rise to 4 to 5 %. The hot metal silicon can be reduced by putting a normal coke rate in the furnace. The "normal" coke rate at "all coke" operation is about 530 kg/tHM. In doing so, however, it takes considerable time to consume all excess coke, which is present in the furnace. More rapid decrease of hot metal silicon can be reached, if a lower coke rate is charged and auxiliary injection is used as soon as required. The injectant is switched on, as soon as the hot metal silicon decreases below 1 %.

An example of such a rapid blow–in of a furnace is presented in Figure 11.7. At the blow–in the furnace was started–up with eight tuyeres (of 36). After opening all tuyeres, a "heavy" burden (coke rate 450 kg/tHM) was put in the furnace 50 hours after the blow–in and coal injection was put on the furnace 58 hours after the blow–in. Hot metal silicon reached the 1% mark 60 hours after the blow–in. The fourth day after the blow–in, average hot metal silicon was 0.95 % and productivity was 2.1 t/m³ WV/d.

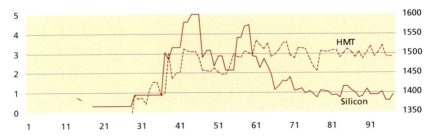

Figure 11.7   Charged coke rate and hot metal silicon after blow–in

Glossary

Angle of repose
The natural angle that is formed when material is discharged onto a pile.

Apatite
A group of phosphate minerals $Ca_5(PO_4)_3(OH, F, Cl)$.

Banded Iron Formation (BIF)
A sedimentary mineral deposit consisting of alternate silica-rich (chert or quartz) and iron rich layers formed 2.5–3.5 billion years ago; the major source of iron ore.

Bentonite
An absorbent aluminum silicate clay formed from volcanic ash and used in various adhesives, cements, and ceramic fillers.

Calcium ferrite
Crystal of CaO and Fe_2O_3.

Chert
A hard, dense sedimentary rock composed of fine-grained silica (SiO_2).

CO_2 Foot Print
The total amount of CO_2 emitted per ton of product over the whole route and taking all energy requirements into account.

Decrepitation
Breaking up of mineral substances when exposed to heat.

Dolomite
Material consisting of lime and magnesium carbonates; extensively used for adjusting the slag composition directly into the blast furnace or via sinter.

Fayalite
Compound of iron silicate: $2FeO.SiO_2$.

Harmonic Mean Size (HMS)
The harmonic mean is the number of values divided by the sum of the reciprocals of the values. This gives a truer average value where ranges of values are used as it tends to mitigate the effect of large outliers from the total data set.

Haematite
Iron oxide in the form of Fe_2O_3.

Magnetite
Iron oxide in the form of Fe_3O_4.

Mill scale
The scale removed in a hot strip mill from the steel slab, mainly iron oxide.

Olivine
A mineral silicate of iron and magnesium, principally $2MgO.SiO_2$, found in igneous and metamorphic rocks and used as a structural material in refractories and in cements.

Serpentine
Any of a group of greenish, brownish, or spotted minerals, $Mg_3Si_2O_5(OH)_4$, used as a source of magnesium and asbestos. Generally a blend of olivine and fayalite with various impurities.

Spinel
Mineral composed of magnesium aluminate.

Wustite
Iron oxide in the form of FeO, does not occur in nature; produced during reduction process.

Annex I *Further Reading*

Babich, A., Senk, D. Gudenau, H.W. Mavrommatis, K. (2008) Ironmaking textbook., RWTH Aachene, Aachen University

Biswas, A.K.: Principles of Blast Furnace Ironmaking, Cootha Publishing House, Brisbane, Australia, 1981.

Committee on Reaction within Blast Furnace, Omori, Y. (chairman): Blast furnace phenomena and modelling, Elsevier, London, 1987.

IISI website: worldsteel.org.

Loison, R., Foch, P., Boyer, A. (1989): Coke quality and production. Butterworths.

McMaster University: Blast Furnace Ironmaking Course (every 2 years), Hamilton, Ontario, Canada, 2006

Meyer, K.: Pelletizing of iron ores, Springer Verlag, Berlin, 1980.

Peacy, J.G. and Davenport, W.G.: The iron blast furnace, Pergamon Press, Oxford, UK, 1979.

Rist, A. and Meysson, N.: A dual graphic representation of the blast–furnace mass and heat balances, Ironmaking proceedings (1966), 88–98.

Rosenqvist, T.: Principles of extractive metallurgy, McGrawHill, Singapore, 1983.

Schoppa, H.: Was der Hochofner von seiner arbeit wissen muss, Verlag Stahleisen, Düsseldorf, Germany, 1992.

Turkdogan, E.T. (1984), Physicochemical aspects of reactions in ironmaking and steelmaking processes, Transactions ISIJ, 24, 591–611.

Wakelin, D.H.: The making, shaping and treating of steel, 11th edition, AISE Steel Foundation, 1999.

Walker, R.D.: Modern Ironmaking Methods, Institute of Metals, London, UK, 1986.

Annex II *References*

Biswas, A.K.: Principles of Blast Furnace Ironmaking, Cootha Publishing House, Brisbane, Australia, 1981.

Bonnekamp, H., Engel, K., Fix, W., Grebe, K. and Winzer, G.: The freezing with nitrogen and dissection of Mannesmann's no 5 blast furnace. Ironmaking proceedings, 1984, Chicago, USA, 139–150.

Carpenter, A. (2006): Use of PCI in blast furnace, IEA Clean coal center

Chaigneau, R., Bakker, T., Steeghs, A. and Bergstrand, R.: Quality assessment of ferrous burden: Utopian dream? 60th Ironmaking Conference Proceedings, 2000, Baltimore, 689–703.

Chaigneau, R.: Complex Calcium Ferrites in the Blast Furnace Process, PhD thesis, Delft University Press, Delft 1994

Committee for Fundamental Metallurgy of the Verein Deutscher Eisenhüttenleute: Slag atlas, Verlag Stahleisen, Düsseldorf, Germany, 1981.

Geerdes, M., Van der Vliet, C., Driessen, J. and Toxopeus, H.: Control of high productivity blast furnace by material distribution, 50th Ironmaking Conference Proceedings, 1991, Vol 50, 367–378.

Grebe, K., Keddeinis, H. and Stricker, K.: Untersuchungen über den Niedrigtemperaturzerfall von Sinter, Stahl und Eisen, 100, (1980), 973–982.

Hartig, W., Langner, K., Lüngen, H.B. and Stricker, K.P.: Measures for increasing the productivity of blast furnace, 59th Ironmaking Conference Proceedings, Pittsburgh, USA, 2000, vol 59, 3–16.

Kolijn, C.: International Cokemaking issues, 3rd McMaster Cokemaking Course, McMaster University, Hamilton, Canada, 2001.

Pagter, J. de and Molenaar, R.: Taphole experience at BF6 and BF7 of Corus Strip Products IJmuiden, McMaster Ironmaking Conference 2001, Hamilton, Canada.

Schoone, E.E., Toxopeus, H. and Vos, D.: Trials with a 100% pellet burden, 54th Ironmaking Conference
Proceedings, Nashville, USA, 1995, vol 54, 465–470.

Singh, B., De, A., Rawat, Y., Das, R. and Chatterjee, A. (1984) Iron and Steel International, Auigust 1984, 135
Slag atlas (1995) Verlag StahlEisen

Toxopeus, H., Steeghs, A. and Van den Boer, J.: PCI at the start of the 21st century, 60th Ironmaking Conference Proceedings, Baltimore, USA, 2001, vol 60, 736–742.

Vander, T., Alvarez, R., Ferraro, M., Fohl, J., Hofherr, K., Huart, J., Mattila, E., Propson, R., Willmers, R. and Van der Velden, B.: Coke quality improvement possibilities and limitations, Proceedings of 3rd International Cokemaking Congress, Gent, Belgium, 1996, vol 3, 28–37.

Annex III *Rules of Thumb*

	Unit	Change	Coke Rate Adj. (kg/t)
Si	%	+ 0.1	+ 4
Moisture	g/m³ STP	+ 10	+ 6
Top pressure	bar	+ 0.1	– 1.2
Coal	kg/t	+ 10	– 9
Oil	kg/t	+ 10	– 11
Oxygen	%	+ 1	+ 1
Blast temperature	°C	+ 100	– 9
Slag	kg/t	+ 10	+ 0.5
Cooling losses	GJ/hr	+ 10	+ 1.2
Gas Utilization	%	+ 1	– 7

Rules of thumb for daily operation of the blast furnace process, a typical example

	Unit	Change	Flame temp. (°C)	Top temp. (°C)
Blast temperature	°C	+ 100	+ 65	– 15
Coal	kg/t	+ 10	– 30	+ 9
Oxygen	%	+ 1	+ 45	– 15
Moisture	g/m³ STP	+ 10	– 50	+ 9

Rules of thumb for daily operation of the blast furnace process (constant blast volume)

Annex IV *Coke Quality Tests*

Since these drum tests are only cold simulations of the load on the coke during its descent through the blast furnace, there are different ideas as to the best way to generate comparative quality values using the drum test. Some of the differences between the various tests are; how the sample is taken as input for the test; the number of rotations; the size of the screens using to determine the size of the resulting coke; and the dimensions of the drum. In Table 1 the differences of the most common used drum tests are presented.

| | Test | | | | | | Strength Indices | |
| | Coke | | Drum | | Test | | Breakage | Abrasion |
	Weight kg	Size mm	Length m	Diam. m	rpm	Total rev.		
Micum	50	> 60	1	1	25	100	M_{40} % > 40 mm	M_{10} % < 10 mm
ISO	50	> 20	1	1	25	100	M_{40} % > 40 mm	M_{10} % < 10 mm
Extended Micum	50	> 60	1	1	25	100, 200, 300, 500, 800	Fissure free size Stabilisation index	Micum Slope
IRSID	50	> 20	1	1	25	500	I_{40} % > 40 mm	I_{10} % < 10 mm
ASTM	10	2–3"	0.46	0.91	24	1,400	% > 1" (25 mm)	% > ¼" (6 mm)
Japanese Drum	10	> 50	1.5	1.5	15	30 or 150		% > 15 mm

Table 1   Differences between the most commonly applied drum tests.

To have a better understanding of coke degradation mechanism under mechanical stress we look at Figure 1. Here the percentage of the coke > 40 mm and < 10 mm of the sample are presented as a function of the number of rotations of the drum.

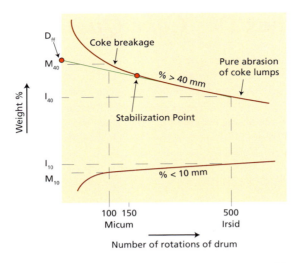

Figure 1   Comparison of different mechanical tumble tests and results.

From this figure we see that the lumps > 40 mm starts to degrade only by breakage until the point of stabilization is reached, when no further breakage occur. From this point on only abrasion takes place to further degrade the coke. In general the coke is stabilized after about 150 rotations of the Micum drum or an equivalent mechanical load. From this figure we see the great difference in number of rotations of the drum between the Micum test and the Irsid test. An advantage of the Irsid test is that the coke is always completely stabilized which makes the result less sensitive for the point of sampling. It further shows that it is in principle not correct to compare test results between different production sites unless the exact the degree of stabilization at the sampling points is known. The weight percentage of coke > 40 mm after 100 rotations is called M_{40} and the percentage after 500 rotations is called the I_{40}. The weight percentage of coke < 10 mm is called M_{10} and I_{10} respectively.

Besides these values, the Fissure Free Size, the Stabilization Index and the Micum slope have been introduced as coke quality parameters. Although in this test the parameter used is not the % > 40 mm of the coke but the average mean size (AMS) as a function of rotations. We will explain these concepts with Figure 1 as well. First we fit a line (shown in green) to the curve of abrasion–only. Then we extrapolate the green line of abrasion–only to the y–intercept (zero rotations) and calculate the AMS of the coke at this point, which gives the Fissure Free Size (FFS), also known as Dff. This then represents the size at which there would be no degradation due to breakage, but only abrasion. The slope of the green line of abrasion–only is called the Micum Slope. Some mills consider this to be a better way to evaluate abradability than traditional M_{10} or I_{10}. The FFS was developed to simulate a maximum obtainable (theoretical) size for stabilized coke. Some believe the FFS approximately represents the size of stabilized industrial coke at the blast furnace stock line, which is then considered a more suitable controlling parameter. Also a stabilization index can be defined as FFS/AMS, which maximum will be 1 for fully stabilized coke.

Chemical reactivity

Besides a high mechanical strength coke should have a high resistance against chemical attack. There are two measurements for the reaction with CO_2 most commonly used, the CRI and the CSR (Coke Reactivity Index and Coke Strength after reaction).

Coke Reactivity Index

Reactivity of coke can be tested in numerous ways, but by far the most common way to determine the coke reactivity is the Nippon Steel Chemical Reactivity Index (CRI). With this test, coke of a certain size is put under a 100% CO_2 atmosphere at 1100°C. The percentage of coke that is gasified after 120 minutes gives the CRI value. The more reactive the coke, the higher the mass loss will be. Reactivity of the coke is mainly determined by the chemical composition of the parent coal blend, because ash components act as catalysts for the reaction of C with CO_2.

Coke Strength after Reaction

Due to the loss of mass whilst under attack by CO_2, the surface layer of the coke particles get very porous and the mechanical strength against abrasion drops rapidly. To measure this effect the reactivity test is normally followed by a tumbler test to determine the residual coke strength. The percentage of particles that remain larger than 10 mm after 600 rotations is called the 'coke strength after reaction' or CSR index. For most coke produced there exists a strong correlation between CRI and CSR.

Before CRI and CSR were developed, a series of relatively expensive tests were carried out under various research projects that involved partially gasifying the coke in its original particle size under realistic blast furnace conditions before subjecting it to the standard drum test. While the results of this costly research work showed exactly how the coke in the blast furnace was subjected to chemical attack, it provided no better information on coke quality than the more–simple method of determining CRI and CSR. These two parameters are now generally adopted by the coke–making industry as the most important parameters for determining coke quality.

Carburization of Hot Metal

There is no standard test for the dissolution of carbon in hot metal, the carburization. Experiments were conducted on this item by the Institute of Ferrous Metallurgy in Germany to compare different cokes of different coal blends and coke making technologies. The experiments showed a very similar behaviour between most cokes. The only exception was the traditionally produced beehive coke. Although it had a very good CSR and CRI it was the only coke examined that cannot be used alone in a blast furnace because of its poor carburization characteristics. Production trials prove that this type of coke can only be used in a mixture with other more reactive coke.

Index

Alkali   144
Angles of repose   79
Apatite   21

Bell less top   7, 79
Belly   3
Bird's nest   126
Blast furnace construction   8
Blow–down   147
Blow–in after reline   148
Blow–in after stop   145
Bosh gas composition   3, 94
Boudouard reaction   97
Burden calculation   59
Burden descent   68
Burden descent, erratic   69, 88
Burden distribution   78
Burden distribution, control scheme   82

Calcium ferrites   27
Carbon and oxygen   96
Carburisation   160
Casthouse, 1 taphole operation   135
Casthouse, dry hearth practice   129, 139
Casting, delayed   133
Casting, no slag   134
Channelling   78
Charging rate   145
Cherts   20
Coal blending   50
Coal injection, coal selection   49
Coal injection, equipment   48
Coal injection, gasification   51
Coal injection, lances   52
Coal injection, oxygen enrichment   52
Coal injection, replacement ratio   50
Cohesive zone, types of   73
Coke   37

Coke layer thickness   85
Coke push   79
Coke quality   39
Coke reactivity   109
Coke size distribution   43
Coke, percentage at wall   84
Coke, analysis   39
Coke, coal blends for   38
Coke, degradation   39
Coke, mechanical strength   44
Coke, quality tests   46, 158
Cold strength   25
Compression (pellets)   31
Counter current reactor   5, 12
CRI   160
CSR   160

Dead man   126
Direct reduction, iron oxides   97
Direct reduction, accompanying elements   98
Double bell top   7, 79

Efficiency   15

Fayalite   29
Fines, in ore burden   77, 142
Flame temperature   95
Fluidisation   78
Forces, vertical   69

Gas composition, vertical distribution   101
Gas flow   71, 76
Gas injection   57
Gas reduction   98
Gas utilisation   15
Gas utilisation, calculation   64
Glossary   151

Hanging   68
Hardgrove index   50
Hearth   3
Hearth, liquid level   131
Heat fluxes   72
Hot blast stoves   6
Hot metal desulphurisation   115
Hot metal, elementary distribution   116
Hot metal, quality   115

Hydrogen, reduction by   102

I10   158
I40   158
Inner volume   9
Instrumentation   90
Iron ore melting   107

Lintel   8
Liquidus temperature   119
Lump ore   34

M10   158
M40   158
Melting capacity of raceway gas   53
Melting zone   see cohesive zone
Mixed layer   79
Moisture input burden   143
Mushroom   125
Nut coke   88

Oil injection   57
Ore burden quality   22
Ore burden, interaction components   35
Ore burden, melting   109
Ore layer thickness   85
Oxygen lancing through taphole   139

Pellet quality   33
Pellet types   31
Permeability   22, 76
Potassium   144
Pressure taps   91
Production rate   94
Productivity, effect metallic iron   107
Productivity, effect oxygen   106
Pulverised coal injection   51

Quenched furnace, reduction progress   101
Quenched furnaces   2

Raceway   11
RAFT   see Flame temperature
Recirculating elements   144
Reducibility   25
Reduction of iron oxides   14
Reduction, by hydrogen   102

Reduction, direct   97
Reduction, gas   98
Reduction–disintegration   22, 25, 108
Residence time, gas
Residence time, ore burden   17
Resistance   see permeability
Rules of thumb, daily operations   157

Segregation   79
Silicon   117
Sinter, effect basicity on structure   27
Sinter, quality   27
Sinter, types   26
Slag basicity, special situations   122
Slag, basicity   119
Slag, composition   118
Slag, primary   121
Slag, properties   119
Slipping   68
Small coke   see nut coke
Sodium   144
Softening–melting   30, 86
Solution loss   98
Soot   51
Spinel   29
Stack   3
Start–up   145
Steelmaking process   112
Stockhouse   6
Stop   145
Sulphur   118
Swelling (pellets)   32
Symmetry, circumferential   56, 112

Taphole   127
Temperature, flame   see flame temperature
Temperature, profile   6, 104
Throat   3
Top gas, calculation of analysis   64
Top gas, formation of   105

Water–gas shift reaction   103
Whisker   32
Working volume   9

Zinc   144